MCAS Subject Test

Mathematics Grade 8

Student Practice Workbook

+ Two Full-Length MCAS Math Tests

Math Notion

www.MathNotion.com

MCAS Subject Test Mathematics Grade 8

Published in the United State of America By

The Math Notion

Web: WWW.MathNotion.com

Email: info@Mathnotion.com

ISBN: 978-1-63620-057-6

The Math Notion

Michael Smith has been a math instructor for over a decade now. He launched the Math Notion. Since 2006, we have devoted our time to both teaching and developing exceptional math learning materials. As a test prep company, we have worked with thousands of students. We have used the feedback of our students to develop a unique study program that can be used by students to drastically improve their math scores fast and effectively. We have more than a thousand Math learning books including:

– SAT Math Prep

– ACT Math Prep

– SSAT/ISEE Math Prep

– Accuplacer Math Prep

– Common Core Math Prep

–many Math Education Workbooks, Study Guides, Practice and Exercise Books

As an experienced Math test preparation company, we have helped many students raise their standardized test scores—and attend the colleges of their dreams: We tutor online and in person, we teach students in large groups, and we provide training materials and textbooks through our website and through Amazon.

You can contact us via email at:

info@Mathnotion.com

Get the Targeted Practice You Need to Ace the MCAS Math Test!

Georgia Milestones Assessment System Subject Test Mathematics Grade 8 includes easy-to-follow instructions, helpful examples, and plenty of math practice problems to assist students to master each concept, brush up their problem-solving skills, and create confidence.

The MCAS math practice book provides numerous opportunities to evaluate basic skills along with abundant remediation and intervention activities. It is a skill that permits you to quickly master intricate information and produce better leads in less time.

Students can boost their test-taking skills by taking the book's two practice MCAS Math exams. All test questions answered and explained in detail.

Important Features of the 8th grade MCAS Math Book:

- A **complete review** of MCAS math test topics,
- Over 2,500 practice problems covering all topics tested,
- The most important concepts you need to know,
- Clear and concise, easy-to-follow sections,
- Well designed for enhanced learning and interest,
- Hands-on experience with all question types,
- **2 full-length practice tests** with detailed answer explanations,
- Cost-Effective Pricing,

Powerful math exercises to help you avoid traps and pacing yourself to beat the Georgia Milestones test. Students will gain valuable experience and raise their confidence by taking 8th grade math practice tests, learning about test structure, and gaining a deeper understanding of what is tested on the MCAS math grade 8. If ever there was a book to respond to the pressure to increase students' test scores, this is it.

WWW.MathNotion.COM

… So Much More Online!

- ✓ FREE Math Lessons
- ✓ More Math Learning Books!
- ✓ Mathematics Worksheets
- ✓ Online Math Tutors

For a PDF Version of This Book

Please Visit WWW.MathNotion.com

Contents

Chapter 1 :
Integers and Number Theory

Topics that you will practice in this chapter:

- ✓ Rounding
- ✓ Whole Number Addition and Subtraction
- ✓ Whole Number Multiplication and Division
- ✓ Rounding and Estimates
- ✓ Adding and Subtracting Integers
- ✓ Multiplying and Dividing Integers
- ✓ Order of Operations
- ✓ Ordering Integers and Numbers
- ✓ Integers and Absolute Value
- ✓ Factoring Numbers
- ✓ Greatest Common Factor (GCF)
- ✓ Least Common Multiple (LCM)

"Wherever there is number, there is beauty." –Proclus

Rounding

Round each number to the nearest ten.

1) 42 = ____
2) 88 = ____
3) 24 = ____
4) 57 = ____
5) 19 = ____
6) 25 = ____
7) 93 = ____
8) 71 = ____
9) 48 = ____
10) 81 = ____
11) 58 = ____
12) 87 = ____

Round each number to the nearest hundred.

13) 198 = ____
14) 387 = ____
15) 816 = ____
16) 101 = ____
17) 321 = ____
18) 433 = ____
19) 579 = ____
20) 825 = ____
21) 580 = ____
22) 868 = ____
23) 480 = ____
24) 287 = ____

Round each number to the nearest thousand.

25) 1,382 = ____
26) 3,420 = ____
27) 4,254 = ____
28) 6,861 = ____
29) 9,099 = ____
30) 22,980 = ____
31) 45,188 = ____
32) 16,808 = ____
33) 52,866 = ____
34) 85,190 = ____
35) 70,990 = ____
36) 26,869 = ____

Rounding and Estimates

Estimate the sum by rounding each number to the nearest ten.

1) $13 + 22 =$ ______

2) $71 + 23 =$ ______

3) $61 + 58 =$ ______

4) $56 + 85 =$ ______

5) $368 + 249 =$ ______

6) $330 + 903 =$ ______

7) $471 + 293 =$ ______

8) $1{,}950 + 2{,}655 =$ ______

Estimate the product by rounding each number to the nearest ten.

9) $32 \times 71 =$ ______

10) $12 \times 33 =$ ______

11) $31 \times 83 =$ ______

12) $19 \times 11 =$ ______

13) $42 \times 76 =$ ______

14) $63 \times 34 =$ ______

15) $19 \times 31 =$ ______

16) $59 \times 71 =$ ______

Estimate the sum or product by rounding each number to the nearest ten.

17) $\begin{array}{r} 29 \\ \times\ 12 \\ \hline \end{array}$ ______

19) $\begin{array}{r} 48 \\ +\ 82 \\ \hline \end{array}$ ______

21) $\begin{array}{r} 37 \\ \times\ 14 \\ \hline \end{array}$ ______

18) $\begin{array}{r} 37 \\ \times\ 26 \\ \hline \end{array}$ ______

20) $\begin{array}{r} 65 \\ +44 \\ \hline \end{array}$ ______

22) $\begin{array}{r} 71 \\ +\ 32 \\ \hline \end{array}$ ______

Adding and Subtracting Integers

Find each sum.

1) $14 + (-6) =$

2) $(-13) + (-20) =$

3) $5 + (-28) =$

4) $50 + (-12) =$

5) $(-7) + (-15) + 3 =$

6) $30 + (-14) + 8 =$

7) $40 + (-10) + (-14) + 17 =$

8) $(-15) + (-20) + 13 + 35 =$

9) $40 + (-20) + (38 - 29) =$

10) $28 + (-12) + (30 - 12) =$

Find each difference.

11) $(-18) - (-7) =$

12) $25 - (-14) =$

13) $(-20) - 36 =$

14) $34 - (-19) =$

15) $51 - (30 - 21) =$

16) $17 - (5) - (-24) =$

17) $(35 + 20) - (-46) =$

18) $48 - 16 - (-8) =$

19) $62 - (28 + 17) - (-15) =$

20) $58 - (-23) - (-31) =$

21) $19 - (-8) - (-13) =$

22) $(19 - 24) - (-14) =$

23) $27 - 33 - (-21) =$

24) $58 - (32 + 24) - (-9) =$

25) $36 - (-30) + (-17) =$

26) $27 - (-42) + (-31) =$

Multiplying and Dividing Integers

Find each product.

1) $(-9) \times (-5) =$

2) $(-3) \times 9 =$

3) $8 \times (-12) =$

4) $(-7) \times (-20) =$

5) $(-3) \times (-5) \times 6 =$

6) $(14 - 3) \times (-8) =$

7) $12 \times (-9) \times (-3) =$

8) $(140 + 10) \times (-2) =$

9) $10 \times (-12 + 8) \times 3 =$

10) $(-8) \times (-5) \times (-10) =$

Find each quotient.

11) $42 \div (-7) =$

12) $(-48) \div (-6) =$

13) $(-40) \div (-8) =$

14) $54 \div (-2) =$

15) $152 \div 19 =$

16) $(-144) \div (-12) =$

17) $180 \div (-10) =$

18) $(-312) \div (-12) =$

19) $221 \div (-13) =$

20) $(-126) \div (6) =$

21) $(-161) \div (-7) =$

22) $-266 \div (-14) =$

23) $(-120) \div (-4) =$

24) $270 \div (-18) =$

25) $(-208) \div (-8) =$

26) $(135) \div (-15) =$

Order of Operations

Evaluate each expression.

1) $7 + (5 \times 4) =$

2) $14 - (3 \times 6) =$

3) $(19 \times 4) + 16 =$

4) $(16 - 7) - (8 \times 2) =$

5) $27 + (18 \div 3) =$

6) $(18 \times 8) \div 6 =$

7) $(32 \div 4) \times (-2) =$

8) $(9 \times 4) + (32 - 18) =$

9) $24 + (4 \times 3) + 7 =$

10) $(36 \times 3) \div (2 + 2) =$

11) $(-7) + (12 \times 3) + 11 =$

12) $(8 \times 5) - (24 \div 6) =$

13) $(7 \times 6 \div 3) - (12 + 9) =$

14) $(13 + 5 - 14) \times 3 - 2 =$

15) $(20 - 14 + 30) \times (64 \div 4) =$

16) $32 + \left(28 - (36 \div 9)\right) =$

17) $(7 + 6 - 4 - 7) + (15 \div 5) =$

18) $(85 - 20) + (20 - 18 + 7) =$

19) $(20 \times 2) + (14 \times 3) - 22 =$

20) $18 + 5 - (30 \times 3) + 20 =$

Ordering Integers and Numbers

Order each set of integers from least to greatest.

1) $8, -10, -5, -3, 4$ ___, ___, ___, ___, ___, ___

2) $-10, -18, 6, 14, 27$ ___, ___, ___, ___, ___, ___

3) $15, -8, -21, 21, -23$ ___, ___, ___, ___, ___, ___

4) $-14, -40, 23, -12, 47$ ___, ___, ___, ___, ___, ___

5) $59, -54, 32, -57, 36$ ___, ___, ___, ___, ___, ___

6) $68, 26, -19, 47, -34$ ___, ___, ___, ___, ___, ___

Order each set of integers from greatest to least.

7) $18, 36, -16, -18, -10$ ___, ___, ___, ___, ___, ___

8) $27, 34, -12, -24, 94$ ___, ___, ___, ___, ___, ___

9) $50, -21, -13, 42, -2$ ___, ___, ___, ___, ___, ___

10) $37, 46, -20, -16, 86$ ___, ___, ___, ___, ___, ___

11) $-18, 88, -26, -59, 75$ ___, ___, ___, ___, ___, ___

12) $-65, -30, -25, 3, 14$ ___, ___, ___, ___, ___, ___

Integers and Absolute Value

Write absolute value of each number.

1) $|-2| =$

2) $|-27| =$

3) $|-20| =$

4) $|14| =$

5) $|6| =$

6) $|-55| =$

7) $|16| =$

8) $|2| =$

9) $|54| =$

10) $|-4| =$

11) $|-11|$

12) $|88| =$

13) $|0| =$

14) $|79| =$

15) $|-32| =$

16) $|-17| =$

17) $|42| =$

18) $|-46| =$

19) $|1| =$

20) $|-40| =$

Evaluate the value.

21) $|-5| - \frac{|-21|}{7} =$

22) $14 - |3 - 15| - |-4| =$

23) $\frac{|-32|}{4} \times |-4| =$

24) $\frac{|7 \times (-3)|}{7} \times \frac{|-19|}{3} =$

25) $|4 \times (-5)| + \frac{|-40|}{5} =$

26) $\frac{|-45|}{9} \times \frac{|-24|}{12} =$

27) $|-12 + 8| \times \frac{|-7 \times 7|}{7}$

28) $\frac{|-11 \times 2|}{4} \times |-16| =$

Factoring Numbers

List all positive factors of each number.

1) 9

2) 16

3) 24

4) 30

5) 26

6) 46

7) 20

8) 68

9) 28

10) 98

11) 14

12) 54

13) 55

14) 18

15) 63

16) 34

17) 50

18) 62

19) 95

20) 64

21) 70

22) 45

23) 22

24) 65

Greatest Common Factor

Find the GCF for each number pair.

1) 6, 2

2) 4, 5

3) 3, 12

4) 7, 3

5) 5, 10

6) 8, 48

7) 6, 18

8) 9, 15

9) 12, 18

10) 4, 36

11) 6, 10

12) 28, 52

13) 25, 10

14) 22, 24

15) 9, 54

16) 8, 54

17) 42, 14

18) 16, 40

19) 9,2, 3

20) 5, 15, 10

21) 7, 9, 2

22) 16, 64

23) 30, 48

24) 36, 63

Least Common Multiple

Find the LCM for each number pair.

1) 6, 9

2) 15, 45

3) 16, 40

4) 12, 36

5) 18, 27

6) 14, 42

7) 6, 30

8) 8, 56

9) 7, 21

10) 8, 20

11) 15, 25

12) 7, 9

13) 4, 11

14) 8, 28

15) 28, 56

16) 40, 50

17) 12, 13

18) 22, 11

19) 36, 20

20) 15, 35

21) 18, 81

22) 30, 54

23) 18,45

24) 75, 25

Answers of Worksheets

Rounding

1) 40	10) 80	19) 600	28) 7,000
2) 90	11) 60	20) 800	29) 9,000
3) 20	12) 90	21) 600	30) 23,000
4) 60	13) 200	22) 900	31) 45,000
5) 20	14) 400	23) 500	32) 17,000
6) 30	15) 800	24) 300	33) 53,000
7) 90	16) 100	25) 1,000	34) 85,000
8) 70	17) 300	26) 3,000	35) 71,000
9) 50	18) 400	27) 4,000	36) 27,000

Rounding and Estimates

1) 30	7) 760	13) 3,200	19) 130
2) 90	8) 4,610	14) 1,800	20) 110
3) 120	9) 2,100	15) 600	21) 400
4) 150	10) 300	16) 4,200	22) 100
5) 620	11) 2,400	17) 300	
6) 1,230	12) 200	18) 1,200	

Adding and Subtracting Integers

1) 8	8) 13	15) 42	22) 9
2) −33	9) 29	16) 36	23) 15
3) −23	10) 34	17) 101	24) 11
4) 38	11) −11	18) 40	25) 49
5) −19	12) 39	19) 32	26) 38
6) 24	13) −56	20) 112	
7) 33	14) 53	21) 40	

Multiplying and Dividing Integers

1) 45	6) −88	11) −6	16) 12
2) −27	7) 324	12) 8	17) −18
3) −96	8) −300	13) 5	18) 26
4) 140	9) −120	14) −27	19) −17
5) 90	10) −400	15) 8	20) −21

21) 23
22) 19
23) 30
24) −15
25) 26
26) −9

Order of Operations

1) 27
2) −4
3) 92
4) −7
5) 33
6) 24
7) −16
8) 50
9) 43
10) 27
11) 40
12) 36
13) −7
14) 10
15) 576
16) 56
17) 5
18) 74
19) 60
20) −47

Ordering Integers and Numbers

1) −10, −5, −3, 4, 8
2) −18, −10, 6, 14, 27
3) −23, −21, −8, 15, 21
4) −40, −14, −12, 23, 47
5) −57, −54, 32, 36, 59
6) −34, −19, 26, 47, 68
7) 36, 18, −10, −16, −18
8) 94, 34, 27, −12, −24
9) 50, 42, −2, −13, −21
10) 86, 46, 37, −16, −20
11) 88, 75, −18, −26, −59
12) 14, 3, −25, −30, −65

Integers and Absolute Value

1) 2
2) 27
3) 20
4) 14
5) 6
6) 55
7) 16
8) 2
9) 54
10) 4
11) 11
12) 88
13) 0
14) 79
15) 32
16) 17
17) 42
18) 46
19) 1
20) 40
21) 2
22) −2
23) 32
24) 19
25) 28
26) 10
27) 28
28) 88

Factoring Numbers

1) 1, 3, 9
2) 1, 2, 4, 8, 16
3) 1, 2, 3, 4, 6, 8, 12, 24
4) 1, 2, 3, 5, 6, 10, 15, 30
5) 1, 2, 13, 26
6) 1, 2, 23, 46
7) 1, 2, 4, 5, 10, 20
8) 1, 2, 4, 17, 34, 68
9) 1, 2, 4, 7, 14, 28
10) 1, 2, 7, 14, 49, 98
11) 1, 2, 7, 14
12) 1, 2, 3, 6, 9, 18, 27, 54
13) 1, 5, 11, 55
14) 1, 2, 3, 6, 9, 18
15) 1, 3, 7, 9, 21, 63
16) 1, 2, 17, 34
17) 1, 2, 5, 10, 25, 50
18) 1, 2, 31, 62
19) 1, 5, 19, 95
20) 1, 2, 4, 8, 16, 32, 64
21) 1, 2, 5, 7, 10, 14, 35, 70
22) 1, 3, 5, 9, 15, 45
23) 1, 2, 11, 22
24) 1, 5, 13, 65

Greatest Common Factor

1) 2	7) 6	13) 5	19) 1
2) 1	8) 3	14) 2	20) 5
3) 3	9) 6	15) 9	21) 1
4) 1	10) 4	16) 2	22) 16
5) 5	11) 2	17) 14	23) 6
6) 8	12) 4	18) 8	24) 9

Least Common Multiple

1) 18	7) 30	13) 44	19) 180
2) 45	8) 56	14) 56	20) 105
3) 80	9) 21	15) 56	21) 162
4) 36	10) 40	16) 200	22) 270
5) 54	11) 75	17) 156	23) 90
6) 42	12) 63	18) 22	24) 75

Chapter 2 :

Fractions and Decimals

Topics that you will practice in this chapter:

- ✓ Simplifying Fractions
- ✓ Adding and Subtracting Fractions
- ✓ Multiplying and Dividing Fractions
- ✓ Adding and Subtract Mixed Numbers
- ✓ Multiplying and Dividing Mixed Numbers
- ✓ Adding and Subtracting Decimals
- ✓ Multiplying and Dividing Decimals
- ✓ Comparing Decimals
- ✓ Rounding Decimals

"A Man is like a fraction whose numerator is what he is and whose denominator is what he thinks of himself. The larger the denominator, the smaller the fraction." –Tolstoy

Simplifying Fractions

Simplify each fraction to its lowest terms.

1) $\frac{5}{10} =$

2) $\frac{28}{35} =$

3) $\frac{27}{36} =$

4) $\frac{40}{80} =$

5) $\frac{14}{56} =$

6) $\frac{32}{48} =$

7) $\frac{52}{65} =$

8) $\frac{15}{60} =$

9) $\frac{80}{160} =$

10) $\frac{55}{77} =$

11) $\frac{28}{112} =$

12) $\frac{32}{64} =$

13) $\frac{63}{72} =$

14) $\frac{81}{90} =$

15) $\frac{35}{105} =$

16) $\frac{25}{70} =$

17) $\frac{80}{280} =$

18) $\frac{12}{81} =$

19) $\frac{36}{186} =$

20) $\frac{240}{540} =$

21) $\frac{70}{560} =$

Find the answer for each problem.

22) Which of the following fractions equal to $\frac{3}{4}$? ____

A. $\frac{60}{90}$ B. $\frac{43}{104}$ C. $\frac{48}{64}$ D. $\frac{150}{300}$

23) Which of the following fractions equal to $\frac{5}{8}$? ____

A. $\frac{125}{200}$ B. $\frac{115}{200}$ C. $\frac{50}{100}$ D. $\frac{30}{90}$

24) Which of the following fractions equal to $\frac{3}{7}$? ____

A. $\frac{58}{116}$ B. $\frac{54}{126}$ C. $\frac{270}{167}$ D. $\frac{42}{63}$

Adding and Subtracting Fractions

Find the sum.

1) $\frac{5}{9}+\frac{4}{9}=$

2) $\frac{1}{2}+\frac{1}{7}=$

3) $\frac{3}{8}+\frac{1}{4}=$

4) $\frac{3}{5}+\frac{1}{2}=$

5) $\frac{1}{4}+\frac{3}{5}=$

6) $\frac{7}{8}+\frac{3}{8}=$

7) $\frac{1}{2}+\frac{7}{10}=$

8) $\frac{2}{5}+\frac{2}{3}=$

9) $\frac{5}{7}+\frac{2}{3}=$

10) $\frac{7}{12}+\frac{3}{4}=$

11) $\frac{5}{6}+\frac{2}{5}=$

12) $\frac{1}{12}+\frac{2}{3}=$

Find the difference.

13) $\frac{1}{3}-\frac{1}{6}=$

14) $\frac{3}{4}-\frac{1}{8}=$

15) $\frac{1}{2}-\frac{1}{3}=$

16) $\frac{1}{4}-\frac{1}{5}=$

17) $\frac{5}{8}-\frac{2}{3}=$

18) $\frac{1}{4}-\frac{1}{7}=$

19) $\frac{5}{6}-\frac{1}{9}=$

20) $\frac{3}{4}-\frac{1}{6}=$

21) $\frac{7}{8}-\frac{1}{12}=$

22) $\frac{8}{15}-\frac{3}{5}=$

23) $\frac{3}{12}-\frac{1}{14}=$

24) $\frac{10}{13}-\frac{7}{26}=$

25) $\frac{6}{7}-\frac{3}{4}=$

26) $\frac{4}{5}-\frac{1}{8}=$

27) $\frac{4}{7}-\frac{2}{35}=$

28) $\frac{9}{16}-\frac{2}{8}=$

29) $\frac{8}{9}-\frac{7}{18}=$

30) $\frac{1}{2}-\frac{4}{9}=$

Multiplying and Dividing Fractions

Find the value of each expression in lowest terms.

1) $\frac{1}{5} \times \frac{15}{5} =$

2) $\frac{9}{12} \times \frac{4}{9} =$

3) $\frac{1}{16} \times \frac{8}{10} =$

4) $\frac{1}{24} \times \frac{8}{10} =$

5) $\frac{1}{5} \times \frac{1}{4} =$

6) $\frac{7}{9} \times \frac{1}{7} =$

7) $\frac{6}{7} \times \frac{1}{3} =$

8) $\frac{2}{8} \times \frac{2}{8} =$

9) $\frac{5}{8} \times \frac{3}{5} =$

10) $\frac{4}{7} \times \frac{1}{8} =$

11) $\frac{7}{15} \times \frac{5}{7} =$

12) $\frac{3}{10} \times \frac{5}{9} =$

Find the value of each expression in lowest terms.

13) $\frac{1}{4} \div \frac{1}{8} =$

14) $\frac{1}{10} \div \frac{1}{5} =$

15) $\frac{3}{4} \div \frac{1}{5} =$

16) $\frac{1}{3} \div \frac{5}{6} =$

17) $\frac{1}{7} \div \frac{8}{42} =$

18) $\frac{3}{4} \div \frac{1}{6} =$

19) $\frac{2}{7} \div \frac{7}{13} =$

20) $\frac{1}{24} \div \frac{3}{16} =$

21) $\frac{7}{12} \div \frac{5}{6} =$

22) $\frac{22}{18} \div \frac{11}{9} =$

23) $\frac{9}{35} \div \frac{3}{7} =$

24) $\frac{2}{7} \div \frac{8}{21} =$

25) $\frac{1}{9} \div \frac{2}{5} =$

26) $\frac{5}{12} \div \frac{3}{5} =$

27) $\frac{3}{20} \div \frac{1}{6} =$

28) $\frac{8}{20} \div \frac{3}{4} =$

29) $\frac{5}{6} \div \frac{2}{9} =$

30) $\frac{5}{11} \div \frac{3}{4} =$

Adding and Subtracting Mixed Numbers

Find the sum.

1) $3\frac{1}{3}+2\frac{1}{6}=$

2) $4\frac{1}{2}+3\frac{1}{2}=$

3) $3\frac{3}{8}+1\frac{1}{8}=$

4) $2\frac{1}{4}+2\frac{1}{3}=$

5) $3\frac{5}{6}+2\frac{7}{12}=$

6) $5\frac{4}{15}+3\frac{3}{5}=$

7) $2\frac{1}{3}+4\frac{3}{7}=$

8) $3\frac{1}{2}+4\frac{2}{5}=$

9) $5\frac{2}{5}+6\frac{3}{7}=$

10) $8\frac{5}{16}+6\frac{1}{12}=$

Find the difference.

11) $3\frac{1}{4}-1\frac{3}{4}=$

12) $6\frac{3}{5}-4\frac{2}{5}=$

13) $4\frac{1}{3}-3\frac{1}{9}=$

14) $7\frac{1}{7}-5\frac{1}{2}=$

15) $5\frac{1}{3}-2\frac{1}{12}=$

16) $8\frac{1}{5}-4\frac{1}{3}=$

17) $9\frac{1}{4}-6\frac{1}{8}=$

18) $11\frac{7}{15}-8\frac{3}{5}=$

19) $14\frac{5}{6}-11\frac{3}{5}=$

20) $18\frac{2}{7}-14\frac{1}{5}=$

21) $9\frac{1}{3}-4\frac{1}{4}=$

22) $6\frac{1}{8}-4\frac{1}{16}=$

23) $19\frac{3}{8}-15\frac{1}{3}=$

24) $11\frac{1}{9}-8\frac{1}{8}=$

25) $17\frac{1}{7}-11\frac{1}{5}=$

26) $16\frac{2}{9}-9\frac{5}{7}=$

Multiplying and Dividing Mixed Numbers

Find the product.

1) $5\frac{1}{2} \times 2\frac{1}{4} =$

2) $5\frac{1}{3} \times 4\frac{1}{3} =$

3) $5\frac{3}{4} \times 6\frac{1}{4} =$

4) $3\frac{1}{3} \times 2\frac{3}{5} =$

5) $4\frac{8}{10} \times 1\frac{1}{24} =$

6) $6\frac{2}{7} \times 1\frac{1}{11} =$

7) $8\frac{2}{3} \times 3\frac{1}{2} =$

8) $3\frac{4}{7} \times 2\frac{1}{5} =$

9) $5\frac{2}{8} \times 4\frac{1}{6} =$

10) $7\frac{3}{3} \times 1\frac{3}{8} =$

Find the quotient.

11) $2\frac{2}{5} \div 4\frac{1}{5} =$

12) $4\frac{1}{6} \div 3\frac{1}{3} =$

13) $6\frac{1}{3} \div 1\frac{1}{2} =$

14) $7\frac{1}{10} \div 2\frac{2}{5} =$

15) $3\frac{1}{3} \div 1\frac{1}{9} =$

16) $1\frac{1}{10} \div 4\frac{1}{2} =$

17) $1\frac{3}{16} \div 5\frac{1}{4} =$

18) $4\frac{1}{3} \div 4\frac{3}{4} =$

19) $9\frac{1}{3} \div 2\frac{1}{4} =$

20) $15\frac{1}{3} \div 5\frac{1}{2} =$

21) $4\frac{1}{6} \div 1\frac{1}{5} =$

22) $1\frac{1}{18} \div 1\frac{2}{9} =$

23) $4\frac{2}{7} \div 1\frac{3}{10} =$

24) $7\frac{1}{3} \div 2\frac{2}{11} =$

25) $8\frac{2}{5} \div 1\frac{1}{6} =$

26) $9\frac{1}{3} \div 2\frac{1}{7} =$

Adding and Subtracting Decimals

Add and subtract decimals.

1) $\begin{array}{r} 35.19 \\ -\ 24.28 \\ \hline \end{array}$ ______

4) $\begin{array}{r} 38.72 \\ -\ 21.68 \\ \hline \end{array}$ ______

7) $\begin{array}{r} 86.09 \\ -\ 35.14 \\ \hline \end{array}$ ______

2) $\begin{array}{r} 34.29 \\ +\ 42.58 \\ \hline \end{array}$ ______

5) $\begin{array}{r} 57.39 \\ +\ 26.54 \\ \hline \end{array}$ ______

8) $\begin{array}{r} 54.51 \\ +\ 32.66 \\ \hline \end{array}$ ______

3) $\begin{array}{r} 61.20 \\ +\ 33.75 \\ \hline \end{array}$ ______

6) $\begin{array}{r} 70.24 \\ -\ 42.35 \\ \hline \end{array}$ ______

9) $\begin{array}{r} 114.21 \\ -\ 88.69 \\ \hline \end{array}$ ______

Find the missing number.

10) ___ + 2.8 = 5.4

11) 4.1 + ___ = 5.88

12) 6.45 + ___ = 8

13) 7.25 − ___ = 3.40

14) ___ − 2.35 = 4.25

15) ___ − 19.85 = 6.54

16) 22.15 + ___ = 28.95

17) ___ − 37.16 = 9.42

18) ___ + 24.50 = 34.19

19) 72.40 + ___ = 125.20

Multiplying and Dividing Decimals

Find the product.

1) $0.5 \times 0.6 =$

2) $3.3 \times 0.4 =$

3) $1.28 \times 0.5 =$

4) $0.35 \times 0.6 =$

5) $1.85 \times 0.6 =$

6) $0.24 \times 0.5 =$

7) $5.25 \times 1.4 =$

8) $18.5 \times 4.6 =$

9) $15.4 \times 6.8 =$

10) $19.5 \times 2.6 =$

11) $32.2 \times 1.5 =$

12) $78.4 \times 4.5 =$

Find the quotient.

13) $1.85 \div 10 =$

14) $74.6 \div 100 =$

15) $3.6 \div 3 =$

16) $9.6 \div 0.4 =$

17) $15.5 \div 0.5 =$

18) $32.8 \div 0.2 =$

19) $22.15 \div 1{,}000 =$

20) $53.55 \div 0.7 =$

21) $322.2 \div 0.2 =$

22) $50.67 \div 0.18 =$

23) $77.4 \div 0.8 =$

24) $27.93 \div 0.03 =$

Comparing Decimals

✍ **Write the correct comparison symbol (>, < or =).**

1) 0.70 □ 0.070

2) 0.049 □ 0.49

3) 5.090 □ 5.09

4) 2.57 □ 2.05

5) 9.03 □ 0.930

6) 6.06 □ 6.6

7) 7.02 □ 7.020

8) 3.04 □ 3.2

9) 3.61 □ 3.245

10) 0.986 □ 0.0986

11) 17.24 □ 17.240

12) 0.759 □ 0.81

13) 9.040 □ 9.40

14) 5.73 □ 5.213

15) 9.44 □ 9.404

16) 7.17 □ 7.170

17) 4.85 □ 4.085

18) 9.041 □ 9.40

19) 3.033 □ 3.030

20) 4.97 □ 4.970

Rounding Decimals

Round each decimal to the nearest whole number.

1) 28.12
2) 6.9
3) 16.22
4) 8.5
5) 7.95
6) 52.7

Round each decimal to the nearest tenth.

7) 31.761
8) 14.421
9) 94.729
10) 77.89
11) 13.219
12) 59.89

Round each decimal to the nearest hundredth.

13) 8.428
14) 23.812
15) 55.3786
16) 231.912
17) 62.241
18) 19.447

Round each decimal to the nearest thousandth.

19) 15.54324
20) 34.62586
21) 243.8652
22) 80.4529
23) 67.1983
24) 72.36788

Answers of Worksheets

Simplifying Fractions

1) $\frac{1}{2}$
2) $\frac{4}{5}$
3) $\frac{3}{4}$
4) $\frac{1}{2}$
5) $\frac{1}{4}$
6) $\frac{2}{3}$
7) $\frac{4}{5}$
8) $\frac{1}{4}$
9) $\frac{1}{2}$
10) $\frac{5}{7}$
11) $\frac{1}{4}$
12) $\frac{1}{2}$
13) $\frac{7}{8}$
14) $\frac{9}{10}$
15) $\frac{1}{3}$
16) $\frac{5}{14}$
17) $\frac{2}{7}$
18) $\frac{4}{27}$
19) $\frac{6}{31}$
20) $\frac{4}{9}$
21) $\frac{1}{8}$
22) C
23) A
24) B

Adding and Subtracting Fractions

1) $\frac{9}{9} = 1$
2) $\frac{9}{14}$
3) $\frac{5}{8}$
4) $1\frac{1}{10}$
5) $\frac{17}{20}$
6) $1\frac{1}{4}$
7) $1\frac{1}{5}$
8) $1\frac{1}{15}$
9) $1\frac{8}{21}$
10) $1\frac{1}{3}$
11) $1\frac{7}{30}$
12) $\frac{3}{4}$
13) $\frac{1}{6}$
14) $\frac{5}{8}$
15) $\frac{1}{6}$
16) $\frac{1}{20}$
17) $-\frac{1}{24}$
18) $\frac{3}{28}$
19) $\frac{13}{18}$
20) $\frac{7}{12}$
21) $\frac{19}{24}$
22) $-\frac{1}{15}$
23) $\frac{5}{28}$
24) $\frac{1}{2}$
25) $\frac{3}{28}$
26) $\frac{27}{40}$
27) $\frac{18}{35}$
28) $\frac{5}{16}$
29) $\frac{1}{2}$
30) $\frac{1}{18}$

Multiplying and Dividing Fractions

1) $\frac{3}{5}$
2) $\frac{1}{3}$
3) $\frac{1}{20}$
4) $\frac{1}{30}$
5) $\frac{1}{20}$
6) $\frac{1}{9}$
7) $\frac{2}{7}$
8) $\frac{1}{16}$
9) $\frac{3}{8}$
10) $\frac{1}{14}$
11) $\frac{1}{3}$
12) $\frac{1}{6}$
13) 2
14) $\frac{1}{2}$
15) $3\frac{3}{4}$
16) $\frac{2}{5}$

17) $\frac{3}{4}$

18) $4\frac{1}{2}$

19) $\frac{26}{49}$

20) $\frac{2}{9}$

21) $\frac{7}{10}$

22) 1

23) $\frac{3}{5}$

24) $\frac{3}{4}$

25) $\frac{5}{18}$

26) $\frac{25}{36}$

27) $\frac{9}{10}$

28) $\frac{8}{15}$

29) $3\frac{3}{4}$

30) $\frac{20}{33}$

Adding and Subtracting Mixed Numbers

1) $5\frac{1}{2}$

2) 8

3) $4\frac{1}{2}$

4) $4\frac{7}{12}$

5) $6\frac{5}{12}$

6) $8\frac{13}{15}$

7) $6\frac{16}{21}$

8) $7\frac{9}{10}$

9) $11\frac{29}{35}$

10) $14\frac{19}{48}$

11) $1\frac{1}{2}$

12) $2\frac{1}{5}$

13) $1\frac{2}{9}$

14) $1\frac{9}{14}$

15) $3\frac{1}{4}$

16) $3\frac{13}{15}$

17) $3\frac{1}{8}$

18) $2\frac{13}{15}$

19) $3\frac{7}{30}$

20) $4\frac{3}{35}$

21) $5\frac{1}{12}$

22) $2\frac{1}{16}$

23) $4\frac{1}{24}$

24) $2\frac{71}{72}$

25) $5\frac{33}{35}$

26) $6\frac{32}{63}$

Multiplying and Dividing Mixed Numbers

1) $12\frac{3}{8}$

2) $23\frac{1}{9}$

3) $35\frac{15}{16}$

4) $8\frac{2}{3}$

5) 5

6) $6\frac{6}{7}$

7) $30\frac{1}{3}$

8) $7\frac{6}{7}$

9) $21\frac{7}{8}$

10) 11

11) $\frac{4}{7}$

12) $1\frac{1}{4}$

13) $4\frac{2}{9}$

14) $2\frac{23}{24}$

15) 3

16) $\frac{11}{45}$

17) $\frac{19}{84}$

18) $\frac{52}{57}$

19) $4\frac{4}{27}$

20) $2\frac{26}{33}$

21) $3\frac{17}{36}$

22) $\frac{19}{22}$

23) $3\frac{27}{91}$

24) $3\frac{13}{36}$

25) $7\frac{1}{5}$

26) $4\frac{16}{45}$

Adding and Subtracting Decimals

1) 10.91

2) 76.87

3) 94.95

4) 17.04

5) 83.93
6) 27.89
7) 50.95
8) 87.17
9) 25.52
10) 2.6
11) 1.78
12) 1.55
13) 3.85
14) 6.6
15) 26.39
16) 6.8
17) 46.58
18) 9.69
19) 52.8

Multiplying and Dividing Decimals

1) 0.3
2) 1.32
3) 0.64
4) 0.21
5) 1.11
6) 0.12
7) 7.35
8) 85.1
9) 104.72
10) 50.7
11) 48.3
12) 352.8
13) 0.185
14) 0.746
15) 1.2
16) 24
17) 31
18) 164
19) 0.02215
20) 76.5
21) 1,611
22) 281.5
23) 96.75
24) 931

Comparing Decimals

1) >
2) <
3) =
4) >
5) >
6) <
7) =
8) <
9) >
10) >
11) =
12) <
13) <
14) >
15) >
16) =
17) >
18) <
19) >
20) =

Rounding Decimals

1) 28
2) 7
3) 16
4) 9
5) 8
6) 53
7) 31.8
8) 14.4
9) 94.7
10) 77.9
11) 13.2
12) 59.9
13) 8.43
14) 23.81
15) 55.38
16) 231.91
17) 62.24
18) 19.45
19) 15.543
20) 34.626
21) 243.865
22) 80.453
23) 67.198
24) 72.368

Chapter 3 :

Proportions, Ratios, and Percent

Topics that you will practice in this chapter:

- ✓ Simplifying Ratios
- ✓ Proportional Ratios
- ✓ Similarity and Ratios
- ✓ Ratio and Rates Word Problems
- ✓ Percentage Calculations
- ✓ Percent Problems
- ✓ Discount, Tax and Tip
- ✓ Percent of Change
- ✓ Simple Interest

Without mathematics, there's nothing you can do. Everything around you is mathematics. Everything around you is numbers." – Shakuntala Devi

Simplifying Ratios

Reduce each ratio.

1) 15: 20 = ___:___
2) 7: 70 = ___:___
3) 16: 28 = ___:___
4) 7: 21 = ___:___
5) 4: 40 = ___:___
6) 6: 48 = ___:___
7) 16: 64 = ___:___
8) 10: 25 = ___:___
9) 8: 48 = ___:___
10) 49: 63 = ___:___
11) 18: 27 = ___:___
12) 35: 10 = ___:___
13) 90: 9 = ___:___
14) 24: 32 = ___:___
15) 7: 56 = ___:___
16) 45: 63 = ___:___
17) 56: 72 = ___:___
18) 26: 13 = ___:___
19) 15: 45 = ___:___
20) 28: 4 = ___:___
21) 24: 48 = ___:___
22) 30: 24 = ___:___
23) 70: 140 = ___:___
24) 6: 180 = ___:___

Write each ratio as a fraction in simplest form.

25) 6: 12 =
26) 30: 50 =
27) 15: 35 =
28) 9: 27 =
29) 8: 24 =
30) 18: 84 =
31) 7: 14 =
32) 7: 35 =
33) 40: 96 =
34) 12: 54 =
35) 44: 52 =
36) 12: 27 =
37) 15: 180 =
38) 39: 143 =
39) 20: 300 =
40) 30: 120 =
41) 56: 42 =
42) 26: 130 =
43) 66: 123 =
44) 70: 630 =
45) 75: 125 =

Proportional Ratios

Fill in the blanks; Calculate each proportion.

1) $3:8 = \_\_ : 48$
2) $2:5 = 20: \_\_$
3) $1:9 = \_\_ : 81$
4) $6:7 = 12: \_\_$
5) $9:2 = 63: \_\_$
6) $8:7 = \_\_ : 49$
7) $20:3 = \_\_ : 15$
8) $1:3 = \_\_ : 75$
9) $7:6 = \_\_ : 60$
10) $8:5 = \_\_ : 45$
11) $3:10 = 60: \_\_$
12) $6:11 = 42: \_\_$

State if each pair of ratios form a proportion.

13) $\frac{3}{20}$ *and* $\frac{9}{60}$
14) $\frac{1}{7}$ *and* $\frac{6}{42}$
15) $\frac{3}{7}$ *and* $\frac{24}{56}$
16) $\frac{4}{9}$ *and* $\frac{12}{18}$
17) $\frac{1}{9}$ *and* $\frac{12}{81}$
18) $\frac{7}{8}$ *and* $\frac{21}{28}$
19) $\frac{9}{13}$ *and* $\frac{27}{39}$
20) $\frac{1}{8}$ *and* $\frac{8}{64}$
21) $\frac{6}{19}$ *and* $\frac{30}{85}$
22) $\frac{5}{9}$ *and* $\frac{40}{81}$
23) $\frac{9}{14}$ *and* $\frac{108}{168}$
24) $\frac{15}{23}$ *and* $\frac{360}{552}$

Calculate each proportion.

25) $\frac{20}{25} = \frac{32}{x}, x =$ ____
26) $\frac{1}{8} = \frac{32}{x}, x =$ ____
27) $\frac{15}{5} = \frac{21}{x}, x =$ ____
28) $\frac{1}{7} = \frac{x}{294}, x =$ ____
29) $\frac{7}{9} = \frac{x}{81}, x =$ ____
30) $\frac{1}{5} = \frac{13}{x}, x =$ ____
31) $\frac{9}{5} = \frac{36}{x}, x =$ ____
32) $\frac{6}{13} = \frac{48}{x}, x =$ ____
33) $\frac{5}{8} = \frac{x}{88}, x =$ ____
34) $\frac{4}{15} = \frac{x}{240}, x =$ ____
35) $\frac{9}{19} = \frac{x}{266}, x =$ ____
36) $\frac{7}{15} = \frac{x}{270}, x =$ ____

Similarity and Ratios

Each pair of figures is similar. Find the missing side.

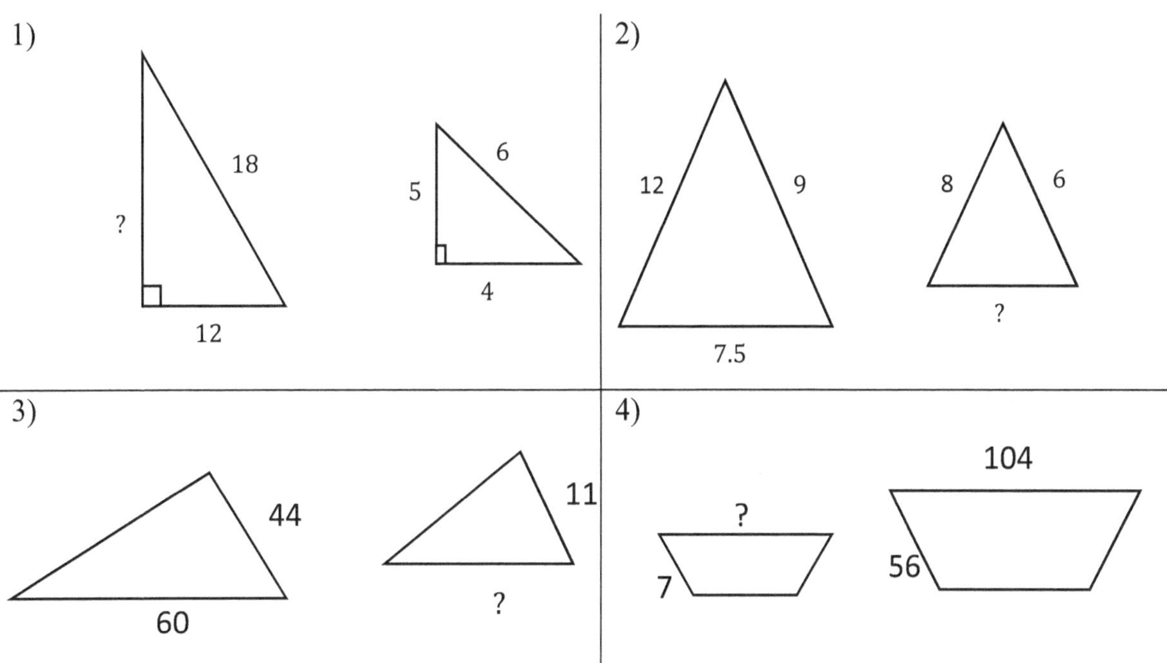

Calculate.

5) Two rectangles are similar. The first is 24 feet wide and 120 feet long. The second is 30 feet wide. What is the length of the second rectangle? ____________

6) Two rectangles are similar. One is 5 meters by 36 meters. The longer side of the second rectangle is 90 meters. What is the other side of the second rectangle? ____________

7) A building casts a shadow 25 ft long. At the same time a girl 10 ft tall casts a shadow 5 ft long. How tall is the building? ____________

8) The scale of a map of Texas is 4 inches: 32 miles. If you measure the distance from Dallas to Martin County as 38.4 inches, approximately how far is Martin County from Dallas? ________

Ratio and Rates Word Problems

Find the answer for each word problem.

1) Mason has 24 red cards and 36 green cards. What is the ratio of Mason 's red cards to his green cards? ______________

2) In a party, 45 soft drinks are required for every 54 guests. If there are 378 guests, how many soft drinks is required? ______________

3) In Mason's class, 42 of the students are tall and 24 are short. In Michael's class 84 students are tall and 48 students are short. Which class has a higher ratio of tall to short students? ______________

4) The price of 5 apples at the Quick Market is $4.6. The price of 7 of the same apples at Walmart is $5.95. Which place is the better buy? ______________

5) The bakers at a Bakery can make 90 bagels in 3 hours. How many bagels can they bake in 24 hours? What is that rate per hour? ______________

6) You can buy 5 cans of green beans at a supermarket for $5.75. How much does it cost to buy 45 cans of green beans? ______________

7) The ratio of boys to girls in a class is 4: 7. If there are 32 boys in the class, how many girls are in that class? ______________

8) The ratio of red marbles to blue marbles in a bag is 3: 7. If there are 50 marbles in the bag, how many of the marbles are red? ______________

Percentage Calculations

Calculate the given percent of each value.

1) $3\%\ of\ 60 =$ ____
2) $20\%\ of\ 32 =$ ____
3) $4\%\ of\ 72 =$ ____
4) $16\%\ of\ 32 =$ ____
5) $25\%\ of\ 124 =$ ____
6) $35\%\ of\ 56 =$ ____
7) $15\%\ of\ 20 =$ ____
8) $14\%\ of\ 150 =$ ____
9) $80\%\ of\ 50 =$ ____
10) $12\%\ of\ 115 =$ ____
11) $72\%\ of\ 250 =$ ____
12) $52\%\ of\ 500 =$ ____
13) $70\%\ of\ 400 =$ ____
14) $27\%\ of\ 145 =$ ____
15) $90\%\ of\ 64 =$ ____
16) $60\%\ of\ 55 =$ ____
17) $22\%\ of\ 210 =$ ____
18) $8\%\ of\ 235 =$ ____

Calculate the percent of each given value.

19) ____$\%\ of\ 25 = 5$
20) ____$\%\ of\ 40 = 20$
21) ____$\%\ of\ 25 = 2$
22) ____$\%\ of\ 50 = 16$
23) ____$\%\ of\ 250 = 5$
24) ____$\%\ of\ 40 = 32$
25) ____$\%\ of\ 125 = 20$
26) ____$\%\ of\ 700 = 49$
27) ____$\%\ of\ 350 = 49$
28) ____% of 500 = 210

Calculate each percent problem.

29) A Cinema has 250 seats. 60 seats were sold for the current movie. What percent of seats are empty? _____ %

30) There are 68 boys and 92 girls in a class. 75% of the students in the class take the bus to school. How many students do not take the bus to school? _____

Percent Problems

Calculate each problem.

1) 9 is what percent of 45? ____%
2) 60 is what percent of 120? ____%
3) 10 is what percent of 200? ____%
4) 15 is what percent of 125? ____%
5) 10 is what percent of 400? ____%
6) 66 is what percent of 55? ____%
7) 40 is what percent of 160? ____%
8) 40 is what percent of 50? ____%
9) 120 is what percent of 800? ____%
10) 78 is what percent of 120? ___%
11) 36 is what percent of 144? ___%
12) 17 is what percent of 85? ___%
13) 90 is what percent of 900? ___%
14) 36 is what percent of 16? ___%
15) 63 is what percent of 14? ___%
16) 18 is what percent of 60? ___%
17) 126 is what percent of 200? ___%
18) 232 is what percent of 40? ___%

Calculate each percent word problem.

19) There are 40 employees in a company. On a certain day, 25 were present. What percent showed up for work? _____%

20) A metal bar weighs 60 ounces. 25% of the bar is gold. How many ounces of gold are in the bar? ___________

21) A crew is made up of 12 women; the rest are men. If 15% of the crew are women, how many people are in the crew? ___________

22) There are 40 students in a class and 8 of them are girls. What percent are boys? _____%

23) The Royals softball team played 400 games and won 280 of them. What percent of the games did they lose? _____%

Discount, Tax and Tip

Find the selling price of each item.

1) Original price of a computer: $420
Tax: 8% Selling price: $______

2) Original price of a laptop: $280
Tax: 4% Selling price: $______

3) Original price of a sofa: $820
Tax: 5% Selling price: $______

4) Original price of a car: $15,800
Tax: 3.6% Selling price: $______

5) Original price of a Table: $250
Tax: 9% Selling price: $______

6) Original price of a house: $630,000
Tax: 1.8% Selling price: $______

7) Original price of a tablet: $450
Discount: 30% Selling price: $____

8) Original price of a chair: $390
Discount: 8% Selling price: $____

9) Original price of a book: $75
Discount: 42% Selling price: $____

10) Original price of a cellphone: $820
Discount: 23% Selling price: $___

11) Food bill: $45
Tip: 15% Price: $______

12) Food bill: $32
Tipp: 20% Price: $______

13) Food bill: $90
Tip: 35% Price: $______

14) Food bill: $42
Tipp: 12% Price: $______

Find the answer for each word problem.

15) Nicolas hired a moving company. The company charged $500 for its services, and Nicolas gives the movers a 40% tip. How much does Nicolas tip the movers? $______

16) Mason has lunch at a restaurant and the cost of his meal is $90. Mason wants to leave a 25% tip. What is Mason's total bill including tip? $_______

17) The sales tax in Texas is 19.80% and an item costs $350. How much is the tax? $_______

18) The price of a table at Best Buy is $680. If the sales tax is 5%, what is the final price of the table including tax? $_______

Percent of Change

Find each percent of change.

1) From 150 to 450. ___%
2) From 50 ft to 250 ft. ___%
3) From $60 to $360. ___%
4) From 60 cm to 180 cm. ___%
5) From 15 to 45. ___%
6) From 80 to 16. ___%
7) From 120 to 360. ___%
8) From 900 to 450. ___%
9) From 1,000 to 200. ___%
10) From 144 to 36. ___%

Calculate each percent of change word problem.

11) Bob got a raise, and his hourly wage increased from $42 to $63. What is the percent increase? ____%

12) The price of a pair of shoes increases from $50 to $61. What is the percent increase? ___%

13) At a coffee shop, the price of a cup of coffee increased from $4.80 to $5.76. What is the percent increase in the cost of the coffee? _____%

14) 51 cm are cut from 85 cm board. What is the percent decrease in length? _____%

15) In a class, the number of students has been increased from 54 to 81. What is the percent increase? _____%

16) The price of gasoline rises from $24.40 to $30.50 in one month. By what percent did the gas price rise? _____%

17) A shirt was originally priced at $38. It went on sale for $24.70. What was the percent that the shirt was discounted? _____%

Simple Interest

Determine the simple interest for these loans.

1) $480 at 11% for 3 years. $ _______
2) $4,200 at 7% for 4 years. $ _______
3) $2,500 at 20% for 3 years. $ _______
4) $6,800 at 3.9% for 4 months. $ ____
5) $800 at 6% for 7 months. $ ______
6) $36,000 at 4.2% for 6 years. $ ____
7) $6,500 at 7% for 4 years. $ ______
8) $850 at 9.5% for 2 years. $ _______
9) $1,200 at 5.8% for 9 months. $ ____
10) $3,000 at 4.5% for 7 years. $ ____

Calculate each simple interest word problem.

11) A new car, valued at $22,000, depreciates at 8.5% per year. What is the value of the car one year after purchase? $___________

12) Sara puts $9,000 into an investment yielding 6% annual simple interest; she left the money in for three years. How much interest does Sara get at the end of those three years? $___________

13) A bank is offering 12% simple interest on a savings account. If you deposit $16,400, how much interest will you earn in two years? $___________

14) $720 interest is earned on a principal of $6,000 at a simple interest rate of 4% interest per year. For how many years was the principal invested? ____

15) In how many years will $2,200 yield an interest of $440 at 4% simple interest? ____

16) Jim invested $8,000 in a bond at a yearly rate of 4.5%. He earned $1,440 in interest. How long was the money invested? ____

Answers of Worksheets

Simplifying Ratios

1) 3: 4
2) 1: 10
3) 4: 7
4) 1: 3
5) 1: 10
6) 1: 8
7) 2: 8
8) 2: 5
9) 1: 6
10) 7: 9
11) 2: 3
12) 7: 2
13) 10: 1
14) 3: 4
15) 1: 8
16) 5: 7
17) 7: 9
18) 2: 1
19) 1: 3
20) 7: 1
21) 1: 2
22) 5: 4
23) 1: 2
24) 1: 30
25) $\frac{1}{2}$
26) $\frac{3}{5}$
27) $\frac{3}{7}$
28) $\frac{1}{3}$
29) $\frac{1}{3}$
30) $\frac{3}{14}$
31) $\frac{1}{2}$
32) $\frac{1}{5}$
33) $\frac{5}{12}$
34) $\frac{2}{9}$
35) $\frac{11}{13}$
36) $\frac{4}{9}$
37) $\frac{1}{12}$
38) $\frac{3}{11}$
39) $\frac{1}{15}$
40) $\frac{1}{4}$
41) $\frac{4}{3}$
42) $\frac{1}{5}$
43) $\frac{22}{41}$
44) $\frac{1}{9}$
45) $\frac{3}{5}$

Proportional Ratios

1) 18
2) 50
3) 9
4) 14
5) 14
6) 56
7) 100
8) 25
9) 70
10) 72
11) 200
12) 77
13) Yes
14) Yes
15) Yes
16) No
17) No
18) No
19) Yes
20) Yes
21) No
22) No
23) Yes
24) Yes
25) 40
26) 256
27) 7
28) 42
29) 63
30) 65
31) 20
32) 104
33) 55
34) 64
35) 126
36) 126

Similarity and ratios

1) 15
2) 5
3) 15
4) 13
5) 150 feet
6) 12.5 meters
7) 50 feet
8) 307.2 miles

Ratio and Rates Word Problems

1) 2: 3
2) 315

3) The ratio for both classes is 7 to 4.
4) Walmart is a better buy.
5) 720, the rate is 30 per hour.
6) $51.75
7) 56
8) 15

Percentage Calculations

1) 1.8
2) 6.4
3) 2.88
4) 5.12
5) 31
6) 19.6
7) 3
8) 21
9) 40
10) 13.8
11) 180
12) 260
13) 280
14) 39.15
15) 57.6
16) 33
17) 46.2
18) 18.8
19) 20%
20) 50%
21) 8%
22) 32%
23) 2%
24) 80%
25) 16%
26) 7%
27) 14%
28) 42%
29) 76%
30) 40

Percent Problems

1) 20%
2) 50%
3) 5%
4) 12%
5) 2.5%
6) 120%
7) 25%
8) 80%
9) 15%
10) 65%
11) 25%
12) 20%
13) 10%
14) 225%
15) 450%
16) 30%
17) 63%
18) 580%
19) 62.5%
20) 15 ounces
21) 80
22) 80%
23) 30%

Discount, Tax and Tip

1) $453.60
2) $291.20
3) $861.00
4) $16,368.80
5) $272.50
6) $641,340
7) $315.00
8) $358.80
9) $43.50
10) $631.40
11) $51.75
12) $38.40
13) $121.50
14) $47.04
15) $200.00
16) $112.50
17) $69.30
18) $714.00

Percent of Change

1) 200%
2) 400%
3) 500%
4) 200%
5) 200%
6) 80%
7) 200%
8) 50%
9) 80%
10) 75%
11) 50%
12) 22%
13) 20%
14) 60%
15) 50%
16) 25%
17) 35%

Simple Interest

1) $158.40
2) $1,176.00
3) $1,500.00
4) $88.40
5) $28.00
6) $9,072.00
7) $1,820.00
8) $161.50
9) $52.20
10) $945.00
11) $20,130.00
12) $1,620.00
13) $3,936.00
14) 3 years
15) 5 years
16) 4 years

Chapter 4 :

Exponents and Radicals Expressions

Topics that you will practice in this chapter:

- ✓ Multiplication Property of Exponents
- ✓ Zero and Negative Exponents
- ✓ Division Property of Exponents
- ✓ Powers of Products and Quotients
- ✓ Negative Exponents and Negative Bases
- ✓ Scientific Notation
- ✓ Square Roots
- ✓ Simplifying Radical Expressions
- ✓ Simplifying Radical Expressions Involving Fractions
- ✓ Multiplying Radical Expressions
- ✓ Adding and Subtracting Radical Expressions

Mathematics is no more computation than typing is literature.

– John Allen Paulos

Multiplication Property of Exponents

Simplify and write the answer in exponential form.

1) $4 \times 4^5 =$

2) $8^4 \times 8 =$

3) $7^3 \times 7^3 =$

4) $9^2 \times 9^2 =$

5) $2^2 \times 2^4 \times 2 =$

6) $5 \times 5^3 \times 5^3 =$

7) $4^3 \times 4^2 \times 4 \times 4 =$

8) $5x \times x =$

9) $x^3 \times x^3 =$

10) $x^7 \times x^2 =$

11) $x^4 \times x^3 \times x^2 =$

12) $10x \times 3x =$

13) $4x^3 \times 4x^3 =$

14) $7x^3 \times x =$

15) $3x^2 \times 4x^2 \times x^2 =$

16) $5x^4 \times x^4 =$

17) $2x^8 \times 2x =$

18) $6x \times x^5 =$

19) $4x^2 \times 6x^6 =$

20) $5yx^3 \times 4x =$

21) $7x^3 \times y^5x^7 =$

22) $y^2x^3 \times y^5x^4 =$

23) $3x^5 \times 4x^3y^4 =$

24) $4x^4 \times 9x^2y^5 =$

25) $5x^3y^4 \times 6x^8y^2 =$

26) $8x^3y^6 \times 4xy^3 =$

27) $2xy^5 \times 6x^3y^3 =$

28) $4x^5y^2 \times 4x^2y^8 =$

29) $7x \times 3y^8x^2 \times y^5 =$

30) $x^3 \times 2y^3x^4 \times 2y =$

31) $3yx^4 \times 3y^4x \times 3xy^3 =$

32) $6y^3 \times 2y^2x^4 \times 10yx^5 =$

Zero and Negative Exponents

Evaluate the following expressions.

1) $1^{-5} =$

2) $4^{-1} =$

3) $0^{10} =$

4) $1^{15} =$

5) $5^{-2} =$

6) $3^{-3} =$

7) $9^{-1} =$

8) $10^{-2} =$

9) $12^{-2} =$

10) $2^{-5} =$

11) $3^{-4} =$

12) $2^{-4} =$

13) $6^{-3} =$

14) $10^{-3} =$

15) $30^{-1=}$

16) $15^{-2} =$

17) $4^{-3} =$

18) $2^{-7} =$

19) $5^{-3} =$

20) $4^{-4} =$

21) $3^{-5} =$

22) $10^{-4} =$

23) $2^{-10} =$

24) $8^{-3} =$

25) $20^{-2} =$

26) $14^{-2} =$

27) $9^{-3} =$

28) $100^{-2} =$

29) $5^{-4} =$

30) $4^{-6} =$

31) $(\frac{1}{4})^{-3}$

32) $(\frac{1}{6})^{-2} =$

33) $(\frac{1}{7})^{-2} =$

34) $(\frac{2}{3})^{-3} =$

35) $(\frac{1}{13})^{-2} =$

36) $(\frac{7}{12})^{-2} =$

37) $(\frac{1}{6})^{-3} =$

38) $(\frac{1}{300})^{-2} =$

39) $(\frac{2}{9})^{-2} =$

40) $(\frac{7}{5})^{-1} =$

41) $(\frac{13}{23})^{0} =$

42) $(\frac{1}{4})^{-5} =$

Division Property of Exponents

Simplify.

1) $\frac{5^6}{5^7} =$

2) $\frac{8^8}{8^6} =$

3) $\frac{4^5}{4} =$

4) $\frac{3}{3^5} =$

5) $\frac{x}{x^6} =$

6) $\frac{3\times3^2}{3^2\times3^5} =$

7) $\frac{9^4}{9^2} =$

8) $\frac{10\times10^9}{10^2\times10^7} =$

9) $\frac{7^5\times7^7}{7^4\times7^8} =$

10) $\frac{15x}{30x^6} =$

11) $\frac{3x^9}{4x^4} =$

12) $\frac{15x^8}{10x^9} =$

13) $\frac{42x^5}{6y^9} =$

14) $\frac{36y^8}{4x^4y^5} =$

15) $\frac{2x^7}{9x} =$

16) $\frac{49x^8y^6}{7x^9} =$

17) $\frac{48x^2}{24x^6y^{12}} =$

18) $\frac{30yx^5}{6yx^7} =$

19) $\frac{19x^7y}{38x^{12}y^4} =$

20) $\frac{9x^8}{63x^8} =$

21) $\frac{9x^{-9}}{4x^{-3}} =$

Powers of Products and Quotients

Simplify.

1) $(4^3)^2 =$

2) $(2^3)^4 =$

3) $(2 \times 2^3)^2 =$

4) $(5 \times 5^5)^6 =$

5) $(19^4 \times 19^2)^3 =$

6) $(2^3 \times 2^4)^4 =$

7) $(5 \times 5^2)^2 =$

8) $(4^4)^4 =$

9) $(8x^5)^2 =$

10) $(3x^2y^4)^4 =$

11) $(7x^5y^2)^2 =$

12) $(5x^4y^4)^3 =$

13) $(2x^3y^3)^5 =$

14) $(10x^3y^4)^3 =$

15) $(13y^3y)^2 =$

16) $(5x^6x^4)^2 =$

17) $(6x^7y^6)^3 =$

18) $(12x^5x^7)^2 =$

19) $(2x^4 \times 2x)^4 =$

20) $(2x^4y^3)^5 =$

21) $(15x^7y^2)^2 =$

22) $(8x^3y^5)^3 =$

23) $(3x \times 2y^2)^4 =$

24) $(\frac{4x}{x^5})^2 =$

25) $\left(\frac{x^4y^5}{x^3y^5}\right)^9 =$

26) $\left(\frac{36xy}{6x^5}\right)^3 =$

27) $\left(\frac{x^7}{x^8y^2}\right)^6 =$

28) $\left(\frac{xy^4}{x^3y^6}\right)^{-3} =$

29) $\left(\frac{5xy^8}{x^3}\right)^2 =$

30) $\left(\frac{xy^6}{2xy^3}\right)^{-4} =$

Negative Exponents and Negative Bases

Simplify.

1) $-9^{-1} =$

2) $-9^{-2} =$

3) $-2^{-5} =$

4) $-x^{-7} =$

5) $11x^{-1} =$

6) $-8x^{-3} =$

7) $-12x^{-5} =$

8) $-9x^{-8}y^{-6} =$

9) $32x^{-5}y^{-1} =$

10) $10a^{-9}b^{-3} =$

11) $-17x^{4}y^{-6} =$

12) $-\frac{25}{x^{-5}} =$

13) $-\frac{13x}{a^{-7}} =$

14) $(-\frac{1}{3})^{-4} =$

15) $(-\frac{3}{4})^{-2} =$

16) $-\frac{14}{a^{-6}b^{-3}} =$

17) $-\frac{7x}{x^{-8}} =$

18) $-\frac{a^{-9}}{b^{-5}} =$

19) $-\frac{11}{x^{-5}} =$

20) $\frac{8b}{-16c^{-6}} =$

21) $\frac{12ab}{a^{-4}b^{-3}} =$

22) $-\frac{8n^{-4}}{32p^{-7}} =$

23) $\frac{16ab^{-6}}{-6c^{-5}} =$

24) $(\frac{10a}{5c})^{-4} =$

25) $(-\frac{12x}{4yz})^{-3} =$

26) $\frac{8ab^{-7}}{-5c^{-3}} =$

27) $(-\frac{x^{4}}{x^{5}})^{-5} =$

28) $(-\frac{x^{-2}}{7x^{3}})^{-2} =$

29) $(-\frac{x^{-4}}{x^{2}})^{-6} =$

Scientific Notation

Write each number in scientific notation.

1) 0.223 =

2) 0.09 =

3) 4.5 =

4) 900 =

5) 2,000 =

6) 0.006 =

7) 33 =

8) 9,400 =

9) 1,470 =

10) 52,000 =

11) 8,000,000 =

12) 0.00009 =

13) 2,158,000 =

14) 0.0039 =

15) 0.000075 =

16) 4,300,000 =

17) 130,000 =

18) 4,000,000,000 =

19) 0.00009 =

20) 0.0039 =

Write each number in standard notation.

21) $4 \times 10^{-1} =$

22) $1.2 \times 10^{-3} =$

23) $2.7 \times 10^{5} =$

24) $6 \times 10^{-4} =$

25) $3.6 \times 10^{-3} =$

26) $5.5 \times 10^{5} =$

27) $3.2 \times 10^{4} =$

28) $3.88 \times 10^{6} =$

29) $7 \times 10^{-6} =$

30) $4.2 \times 10^{-7} =$

Square Roots

Find the value each square root.

1) $\sqrt{16}$ = ____
2) $\sqrt{25}$ = ____
3) $\sqrt{1}$ = ____
4) $\sqrt{64}$ = ____
5) $\sqrt{0}$ = ____
6) $\sqrt{196}$ = ____
7) $\sqrt{4}$ = ____
8) $\sqrt{256}$ = ____
9) $\sqrt{36}$ = ____
10) $\sqrt{289}$ = ____
11) $\sqrt{169}$ = ____
12) $\sqrt{144}$ = ____
13) $\sqrt{100}$ = ____
14) $\sqrt{1{,}600}$ = ____
15) $\sqrt{2{,}500}$ = ____
16) $\sqrt{324}$ = ____
17) $\sqrt{529}$ = ____
18) $\sqrt{20}$ = ____
19) $\sqrt{625}$ = ____
20) $\sqrt{18}$ = ____
21) $\sqrt{50}$ = ____
22) $\sqrt{1{,}024}$ = ____
23) $\sqrt{160}$ = ____
24) $\sqrt{32}$ = ____

Evaluate.

25) $\sqrt{4} \times \sqrt{25}$ = ________
26) $\sqrt{36} \times \sqrt{49}$ = ________
27) $\sqrt{6} \times \sqrt{6}$ = ________
28) $\sqrt{13} \times \sqrt{13}$ = ________
29) $2\sqrt{5} \times 3\sqrt{5}$ = ________
30) $\sqrt{12} \times \sqrt{3}$ = ________
31) $\sqrt{13} + \sqrt{13}$ = ________
32) $\sqrt{10} + 2\sqrt{10}$ = ________
33) $12\sqrt{7} - 10\sqrt{7}$ = ________
34) $4\sqrt{10} \times 2\sqrt{10}$ = ________
35) $5\sqrt{3} \times 8\sqrt{3}$ = ________
36) $6\sqrt{3} - \sqrt{12}$ = ________

Simplifying Radical Expressions

Simplify.

1) $\sqrt{13x^2} =$

2) $\sqrt{75x^2} =$

3) $\sqrt[3]{27a} =$

4) $\sqrt{64x^5} =$

5) $\sqrt{216a} =$

6) $\sqrt[3]{63w^3} =$

7) $\sqrt{192x} =$

8) $\sqrt{125v} =$

9) $\sqrt[3]{128x^2} =$

10) $\sqrt{100x^9} =$

11) $\sqrt{16x^4} =$

12) $\sqrt[3]{500a^5} =$

13) $\sqrt{242} =$

14) $\sqrt{392p^3} =$

15) $\sqrt{8m^6} =$

16) $\sqrt{198x^3y^3} =$

17) $\sqrt{121x^5y^5} =$

18) $\sqrt{16a^6b^3} =$

19) $\sqrt{90x^5y^7} =$

20) $\sqrt[3]{64y^2x^6} =$

21) $10\sqrt{16x^4} =$

22) $6\sqrt{81x^2} =$

23) $\sqrt[3]{56x^2y^6} =$

24) $\sqrt[3]{1{,}000x^5y^7} =$

25) $8\sqrt{50a} =$

26) $\sqrt[4]{625x^8y} =$

27) $\sqrt{24x^4y^5r^3} =$

28) $5\sqrt{36x^4y^5z^8} =$

29) $3\sqrt[3]{343x^9y^7} =$

30) $5\sqrt{81a^5b^2c^9} =$

31) $\sqrt[4]{625x^8y^{16}} =$

Multiplying Radical Expressions

Simplify.

1) $\sqrt{5} \times \sqrt{5} =$

2) $\sqrt{5} \times \sqrt{10} =$

3) $\sqrt{3} \times \sqrt{12} =$

4) $\sqrt{49} \times \sqrt{47} =$

5) $\sqrt{7} \times -2\sqrt{28} =$

6) $3\sqrt{15} \times \sqrt{5} =$

7) $4\sqrt{72} \times \sqrt{2} =$

8) $\sqrt{5} \times -\sqrt{49} =$

9) $\sqrt{55} \times \sqrt{11} =$

10) $7\sqrt{42} \times 2\sqrt{216} =$

11) $\sqrt{45}(5 + \sqrt{5}) =$

12) $\sqrt{13x^2} \times \sqrt{13x^3} =$

13) $-2\sqrt{27} \times \sqrt{3} =$

14) $2\sqrt{13x^4} \times \sqrt{13x^4} =$

15) $\sqrt{14x^3} \times \sqrt{7x^2} =$

16) $-8\sqrt{5x} \times \sqrt{7x^5} =$

17) $-2\sqrt{16x^5} \times 4\sqrt{8x^3} =$

18) $-4\sqrt{32}(8 + \sqrt{32}) =$

19) $\sqrt{32x}\,(10 - \sqrt{2x}) =$

20) $\sqrt{2x}(8\sqrt{x^5} + \sqrt{8}) =$

21) $\sqrt{20r}\,(5 + \sqrt{5}) =$

22) $-4\sqrt{7x} \times 3\sqrt{14x^5} =$

23) $-2\sqrt{12x} \times 3\sqrt{2x}$

24) $-\sqrt{7v^3}\,(-3\sqrt{42v}) =$

25) $(\sqrt{11} - 5)(\sqrt{11} + 5) =$

26) $(-3\sqrt{5} + 3)(\sqrt{5} - 4) =$

27) $(4 - 6\sqrt{3})(-6 + \sqrt{3}) =$

28) $(8 - 3\sqrt{5})(7 - \sqrt{5}) =$

29) $(-1 - \sqrt{3x})(4 + \sqrt{3x}) =$

30) $(-5 + 2\sqrt{7r})(-5 + \sqrt{7r}) =$

31) $(-\sqrt{7n} + 1)(-\sqrt{7} - 5) =$

32) $(-3 + \sqrt{3})(5 - 2\sqrt{3x}) =$

Simplifying Radical Expressions Involving Fractions

Simplify.

1) $\frac{\sqrt{5}}{\sqrt{3}} =$

2) $\frac{\sqrt{18}}{\sqrt{45}} =$

3) $\frac{\sqrt{10}}{5\sqrt{2}} =$

4) $\frac{13}{\sqrt{3}} =$

5) $\frac{12\sqrt{5r}}{\sqrt{m^5}} =$

6) $\frac{11\sqrt{2}}{\sqrt{k}} =$

7) $\frac{6\sqrt{20x^3}}{\sqrt{16x}} =$

8) $\frac{\sqrt{14x^3y^4}}{\sqrt{7x^4y^3}} =$

9) $\frac{1}{1-\sqrt{5}} =$

10) $\frac{1-8\sqrt{a}}{\sqrt{11a}} =$

11) $\frac{\sqrt{a}}{\sqrt{a}+\sqrt{b}} =$

12) $\frac{1-\sqrt{5}}{2-\sqrt{6}} =$

13) $\frac{4+\sqrt{7}}{3-\sqrt{8}} =$

14) $\frac{5}{-3-3\sqrt{3}}$

15) $\frac{7}{2-\sqrt{5}} =$

16) $\frac{\sqrt{7}-\sqrt{3}}{\sqrt{3}-\sqrt{7}} =$

17) $\frac{\sqrt{5}+\sqrt{7}}{\sqrt{7}-\sqrt{5}} =$

18) $\frac{2\sqrt{2}-\sqrt{3}}{3\sqrt{2}+\sqrt{5}} =$

19) $\frac{\sqrt{11}+5\sqrt{3}}{4-\sqrt{11}} =$

20) $\frac{\sqrt{5}+\sqrt{3}}{2-\sqrt{3}} =$

21) $\frac{\sqrt{32a^7b^4}}{\sqrt{2ab^3}} =$

22) $\frac{10\sqrt{21x^5}}{5\sqrt{x^3}} =$

Adding and Subtracting Radical Expressions

Simplify.

1) $\sqrt{2}+\sqrt{8}=$

2) $3\sqrt{50}+4\sqrt{2}=$

3) $2\sqrt{12}-4\sqrt{3}=$

4) $5\sqrt{32}-5\sqrt{2}=$

5) $3\sqrt{75}-5\sqrt{3}=$

6) $-\sqrt{72}-4\sqrt{2}=$

7) $-7\sqrt{16}-4\sqrt{25}=$

8) $8\sqrt{24}+2\sqrt{6}=$

9) $10\sqrt{49}-7\sqrt{100}=$

10) $-7\sqrt{5}+9\sqrt{45}=$

11) $-15\sqrt{12}+14\sqrt{48}=$

12) $20\sqrt{4}-2\sqrt{25}=$

13) $-2\sqrt{20}+7\sqrt{5}=$

14) $8\sqrt{7}-2\sqrt{63}=$

15) $5\sqrt{44}+3\sqrt{11}=$

16) $3\sqrt{27}-5\sqrt{48}=$

17) $\sqrt{144}-\sqrt{81}=$

18) $3\sqrt{20}-6\sqrt{5}=$

19) $-2\sqrt{7}+8\sqrt{28}=$

20) $3\sqrt{75}-2\sqrt{3}=$

21) $5\sqrt{27}-3\sqrt{3}=$

22) $-7\sqrt{30}+6\sqrt{120}=$

23) $-7\sqrt{24}-2\sqrt{6}=$

24) $-\sqrt{32x}+4\sqrt{2x}=$

25) $\sqrt{7y^2}+y\sqrt{112}=$

26) $\sqrt{45mn^2}+2n\sqrt{5m}=$

27) $-4\sqrt{12a}-4\sqrt{3a}=$

28) $-5\sqrt{15ab}-2\sqrt{60ab}=$

29) $\sqrt{45x^2y}+x\sqrt{20y}=$

30) $2\sqrt{7\text{a}}+4\sqrt{63\text{a}}=$

Answers of Worksheets

Multiplication Property of Exponents

1) 4^6
2) 8^5
3) 7^6
4) 9^4
5) 2^7
6) 5^7
7) 4^7
8) $5x^2$
9) x^6
10) x^9
11) x^9
12) $30x^2$
13) $16x^6$
14) $7x^4$
15) $12x^6$
16) $5x^8$
17) $4x^9$
18) $6x^6$
19) $24x^8$
20) $20x^4y$
21) $7x^{10}y^5$
22) x^7y^7
23) $12x^8y^4$
24) $36x^6y^5$
25) $30x^{11}y^6$
26) $32x^4y^9$
27) $12x^4y^8$
28) $16x^7y^{10}$
29) $21x^3y^{13}$
30) $4x^7y^4$
31) $27x^6y^8$
32) $120x^9y^6$

Zero and Negative Exponents

1) 1
2) $\frac{1}{4}$
3) 0
4) 1
5) $\frac{1}{25}$
6) $\frac{1}{27}$
7) $\frac{1}{9}$
8) $\frac{1}{100}$
9) $\frac{1}{144}$
10) $\frac{1}{32}$
11) $\frac{1}{81}$
12) $\frac{1}{16}$
13) $\frac{1}{216}$
14) $\frac{1}{1,000}$
15) $\frac{1}{30}$
16) $\frac{1}{225}$
17) $\frac{1}{64}$
18) $\frac{1}{128}$
19) $\frac{1}{125}$
20) $\frac{1}{256}$
21) $\frac{1}{243}$
22) $\frac{1}{10,000}$
23) $\frac{1}{1,024}$
24) $\frac{1}{512}$
25) $\frac{1}{400}$
26) $\frac{1}{196}$
27) $\frac{1}{729}$
28) $\frac{1}{10,000}$
29) $\frac{1}{625}$
30) $\frac{1}{4,096}$
31) 64
32) 36
33) 49
34) $\frac{27}{8}$
35) 169
36) $\frac{144}{49}$
37) 216
38) 90,000
39) $\frac{81}{4}$
40) $\frac{5}{7}$
41) 1
42) 1,024

Division Property of Exponents

1) $\frac{1}{5}$
2) 8^2
3) 4^4
4) $\frac{1}{3^4}$
5) $\frac{1}{x^5}$
6) $\frac{1}{3^4}$
7) 9^2
8) 10
9) 1
10) $\frac{1}{2x^5}$
11) $\frac{3x^5}{4}$
12) $\frac{3}{2x}$
13) $\frac{7x^5}{y^9}$
14) $\frac{9y^3}{x^4}$
15) $\frac{2x^6}{9}$

16) $\frac{7y^6}{x}$

17) $\frac{2}{x^4y^{12}}$

18) $\frac{5}{x^2}$

19) $\frac{1}{2x^5y^3}$

20) $\frac{1}{7}$

21) $\frac{9}{4x^6}$

Powers of Products and Quotients

1) 4^6
2) 2^{12}
3) 2^8
4) 5^{36}
5) 19^{18}
6) 2^{28}
7) 5^6
8) 4^{16}
9) $64x^{10}$
10) $81x^8y^{16}$
11) $49x^{10}y^4$
12) $125x^{12}y^{12}$
13) $32x^{15}y^{15}$
14) $1{,}000x^9y^{12}$
15) $169y^8$
16) $25x^{20}$
17) $216x^{21}y^{18}$
18) $144x^{24}$
19) $256x^{20}$
20) $32x^{20}y^{15}$
21) $225x^{14}y^4$
22) $512x^9y^{15}$
23) $1{,}296x^4y^8$
24) $\frac{16}{x^8}$
25) x^9
26) $\frac{216y^3}{x^{12}}$
27) $\frac{1}{x^6y^{12}}$
28) x^6y^6
29) $\frac{25y^{16}}{x^4}$
30) $\frac{16}{y^{12}}$

Negative Exponents and Negative Bases

1) $-\frac{1}{9}$
2) $-\frac{1}{81}$
3) $-\frac{1}{32}$
4) $-\frac{1}{x^7}$
5) $\frac{11}{x}$
6) $-\frac{8}{x^3}$
7) $-\frac{12}{x^5}$
8) $-\frac{9}{x^8y^6}$
9) $\frac{32}{x^5y}$
10) $\frac{10}{a^9b^3}$
11) $-\frac{17x^4}{y^6}$
12) $-25x^5$
13) $-13xa^7$
14) 81
15) $\frac{16}{9}$
16) $-14a^6b^3$
17) $-7x^9$
18) $-\frac{b^5}{a^9}$
19) $-11x^5$
20) $-\frac{bc^6}{2}$
21) $12a^5b^4$
22) $-\frac{p^7}{4n^4}$
23) $-\frac{8ac^5}{3b^6}$
24) $\frac{c^4}{16a^4}$
25) $\frac{y^3z^3}{27x^3}$
26) $-\frac{8ac^3}{5b^7}$
27) $-x^5$
28) $49x^{10}$
29) x^{36}

Scientific Notation

1) 2.23×10^{-1}
2) 9×10^{-2}
3) 4.5×10^{0}
4) 9×10^{2}
5) 2×10^{3}
6) 6×10^{-3}
7) 3.3×10^{1}
8) 9.4×10^{3}
9) 1.47×10^{3}
10) 5.2×10^{4}
11) 8×10^{6}
12) 9×10^{-5}
13) 2.158×10^{6}
14) 3.9×10^{-3}
15) 7.5×10^{-5}
16) 4.3×10^{6}
17) 1.3×10^{5}
18) 4×10^{9}
19) 9×10^{-5}
20) 3.9×10^{-3}
21) 0.4
22) 0.0012
23) 270,000
24) 0.0006
25) 0.0036
26) 550,000
27) 32,000
28) 3,880,000
29) 0.000007
30) 0.00000042

Square Roots

1) 4
2) 5
3) 1
4) 8
5) 0
6) 14
7) 2
8) 16
9) 6
10) 17
11) 13
12) 12
13) 10
14) 40
15) 50
16) 18
17) 23
18) $2\sqrt{5}$
19) 25
20) $3\sqrt{2}$
21) $5\sqrt{2}$
22) 32
23) $4\sqrt{10}$
24) $4\sqrt{2}$
25) 10
26) 42
27) 6
28) 13
29) 30
30) 6
31) $2\sqrt{13}$
32) $3\sqrt{10}$
33) $2\sqrt{7}$
34) 80
35) 120
36) $4\sqrt{3}$

Simplifying radical expressions

1) $x\sqrt{13}$
2) $5x\sqrt{3}$
3) $3\sqrt[3]{a}$
4) $8x^2\sqrt{x}$
5) $6\sqrt{6a}$
6) $w\sqrt[3]{63}$
7) $8\sqrt{3x}$
8) $5\sqrt{5v}$
9) $4\sqrt[3]{2x^2}$
10) $10x^4\sqrt{x}$
11) $4x^2$
12) $5a\sqrt[3]{4a^2}$
13) $11\sqrt{2}$
14) $14p\sqrt{2p}$
15) $2m^3\sqrt{2}$
16) $3x.y\sqrt{22xy}$
17) $11x^2y^2\sqrt{xy}$
18) $4a^3b\sqrt{b}$
19) $3x^2y^3\sqrt{10xy}$
20) $4x^2\sqrt[3]{y^2}$
21) $40x^2$
22) $54x$
23) $2y^2\sqrt[3]{7x^2}$
24) $10xy^2\sqrt[3]{x^2y}$

25) $40\sqrt{2a}$

26) $5x^2\sqrt[4]{y}$

27) $2x^2y^2r\sqrt{6yr}$

28) $30x^2y^2z^4\sqrt{y}$

29) $21x^3y^2\sqrt[3]{y}$

30) $45a^2bc^4\sqrt{ac}$

31) $5x^2y^4$

Multiplying radical expressions

1) 5
2) $5\sqrt{2}$
3) 6
4) $7\sqrt{47}$
5) -28
6) $15\sqrt{3}$
7) 48
8) $-5\sqrt{7}$
9) $11\sqrt{5}$
10) $504\sqrt{7}$
11) $15\sqrt{5}+15$
12) $13x^2\sqrt{x}$
13) -18
14) $26x^4$
15) $7x^2\sqrt{2x}$
16) $-8x^3\sqrt{35}$
17) $-64x^4\sqrt{2}$
18) $-128\sqrt{2}-128$
19) $40\sqrt{2x}-8x$
20) $8x^3\sqrt{2}+4\sqrt{x}$
21) $10\sqrt{5r}+10\sqrt{r}$
22) $-84x^3\sqrt{2}$
23) $-12\sqrt{6x}$
24) $21v^2\sqrt{6}$
25) -14
26) $15\sqrt{5}-27$
27) $40\sqrt{3}-42$
28) $71-29\sqrt{5}$
29) $-3x-5\sqrt{3x}-4$
30) $14r-15\sqrt{7r}+25$
31) $7\sqrt{n}+5\sqrt{7n}-\sqrt{7}-5$
32) $-15+6\sqrt{3x}+5\sqrt{3}-6\sqrt{x}$

Simplifying radical expressions involving fractions

1) $\frac{\sqrt{15}}{3}$
2) $\frac{9\sqrt{10}}{45}=\frac{\sqrt{10}}{5}$
3) $\frac{\sqrt{20}}{10}=\frac{\sqrt{5}}{5}$
4) $\frac{13\sqrt{3}}{3}$
5) $\frac{12\sqrt{5mr}}{m^3}$
6) $\frac{11\sqrt{2k}}{k}$
7) $3x\sqrt{5}$
8) $\frac{\sqrt{2x}}{xy}$
9) $\frac{-1-\sqrt{5}}{4}$
10) $\frac{\sqrt{11a}-8a\sqrt{11}}{11a}$

11) $\frac{a-\sqrt{ab}}{a-b}$

12) $\frac{\sqrt{30}+2\sqrt{5}-\sqrt{6}-2}{2}$

13) $12+8\sqrt{2}+3\sqrt{7}+2\sqrt{14}$

14) $-\frac{5(\sqrt{3}-1)}{6}$

15) $-14-7\sqrt{5}$

16) -1

17) $6+\sqrt{35}$

18) $\frac{12-2\sqrt{10}-3\sqrt{6}+\sqrt{15}}{13}$

19) $\frac{4\sqrt{11}+11+20\sqrt{3}+5\sqrt{33}}{5}$

20) $2\sqrt{5}+3+\sqrt{15}+2\sqrt{3}$

21) $4a^3\sqrt{b}$

22) $2x\sqrt{21}$

Adding and subtracting radical expressions

1) $3\sqrt{2}$

2) $19\sqrt{2}$

3) 0

4) $15\sqrt{2}$

5) $10\sqrt{3}$

6) $-10\sqrt{2}$

7) -48

8) $18\sqrt{6}$

9) 0

10) $20\sqrt{5}$

11) $26\sqrt{3}$

12) 30

13) $3\sqrt{5}$

14) $2\sqrt{7}$

15) $13\sqrt{11}$

16) $-11\sqrt{3}$

17) 3

18) 0

19) $14\sqrt{7}$

20) $13\sqrt{3}$

21) $12\sqrt{3}$

22) $5\sqrt{30}$

23) $-16\sqrt{6}$

24) 0

25) $5y\sqrt{7}$

26) $5n\sqrt{5m}$

27) $-12\sqrt{3a}$

28) $-9\sqrt{15ab}$

29) $5x\sqrt{5y}$

30) $14\sqrt{7a}$

Chapter 5 :
Algebraic Expressions

Topics that you will practice in this chapter:

- ✓ Simplifying Variable Expressions
- ✓ Simplifying Polynomial Expressions
- ✓ Translate Phrases into an Algebraic Statement
- ✓ The Distributive Property
- ✓ Evaluating One Variable Expressions
- ✓ Evaluating Two Variables Expressions
- ✓ Combining like Terms

Mathematics is, as it were, a sensuous logic, and relates to philosophy as do the arts, music, and plastic art to poetry. — K. Shegel

Simplifying Variable Expressions

Simplify each expression.

1) $3(x+5) =$

2) $(-4)(7x-5) =$

3) $11x+5-6x =$

4) $-4-2x^2-6x^2 =$

5) $7+13x^2+3 =$

6) $3x^2+7x+15x^2 =$

7) $3x^2-12x^2+4x =$

8) $4x^2-8x-2x =$

9) $6x+7(3-4x) =$

10) $8x+4(15x-3) =$

11) $6(-3x-9)-17 =$

12) $-11x^2-(-5x) =$

13) $2x+7+5-8x =$

14) $7+6x-11-5x =$

15) $27x+8-13-5x =$

16) $(-11)(-5x+2)-41x =$

17) $19x-4(4-2x) =$

18) $16x+3(3x+6)+10 =$

19) $5(-2x-4)-13x =$

20) $16x-3x(x+10) =$

21) $17x+5x(2-4x) =$

22) $5x(-4x-7)+20x =$

23) $25x-19+4x^2 =$

24) $6x(x-11)+25 =$

25) $4x-5+15x+3x^2 =$

26) $-7x^2-11x-9x =$

27) $10x-9x^2-3x^2-7 =$

28) $13+3x^2-9x^2-21x =$

29) $22x+10x^2-15x+17 =$

30) $4x^2+25x+21x^2 =$

31) $29-12x^2-23x-4x^2 =$

32) $22x-19x-9x^2+30 =$

Simplifying Polynomial Expressions

Simplify each polynomial.

1) $(2x^3 + 8x^2) - (11x + 3x^2) =$ ____________________

2) $(2x^5 + 7x^3) - (5x^3 + 11x^2) =$ ____________________

3) $(41x^4 + 5x^2) - (4x^2 + 20x^4) =$ ____________________

4) $13x - 8x^2 + 4(4x^2 + 3x^3) =$ ____________________

5) $(4x^3 - 22) + 5(3x^2 - 6x^3) =$ ____________________

6) $(4x^3 - 3x) - 5(2x^3 + x^4) =$ ____________________

7) $5(5x - 2x^3) - 2(8x^3 + 5x^2) =$ ____________________

8) $(3x^2 - 10x) - (5x^3 + 14x^2) =$ ____________________

9) $5x^3 - (3x^4 + 5x) + 2x^2 =$ ____________________

10) $11x^4 - (3x^2 + 5x) + 7x =$ ____________________

11) $(6x^2 - 3x^4) - (10x^4 + 3x^2) =$ ____________________

12) $2x^2 - 7x^3 + 19x^4 - 22x^3 =$ ____________________

13) $10x^2 - x^4 + 4x^4 - 32x^3 =$ ____________________

14) $-5x^2 + 17x^3 - 8x^2 - 6x =$ ____________________

15) $x^4 - 11x^5 - 30x^4 + 5x^2 =$ ____________________

16) $21x^3 + 13x - 5x^2 - 11x^3 =$ ____________________

Translate Phrases into an Algebraic Statement

Write an algebraic expression for each phrase.

1) 9 multiplied by x. ________________

2) Subtract 11 from y. ________________

3) 19 divided by x. ________________

4) 38 decreased by y. ________________

5) Add y to 40. ________________

6) The square of 6. ________________

7) x raised to the fifth power. ________________

8) The sum of six and a number. ________________

9) The difference between fifty–seven and y. ________________

10) The quotient of nine and a number. ________________

11) The quotient of the square of x and 25. ________________

12) The difference between x and 6 is 19. ________________

13) 10 times a reduced by the square of b. ________________

14) Subtract the product of a and b from 41. ________________

The Distributive Property

Use the distributive property to simply each expression.

1) $4(1+2x)=$

2) $2(4+7x)=$

3) $3(4x-4)=$

4) $(2x-5)(-6)=$

5) $(-3)(x+6)=$

6) $(4+3x)2=$

7) $(-5)(8-3x)=$

8) $-(-5-7x)=$

9) $(-6x+3)(-3)=$

10) $(-4)(x-7)=$

11) $-(5-3x)=$

12) $3(9+4x)=$

13) $6(4+3x)=$

14) $(-5x+3)2=$

15) $(5-8x)(-3)=$

16) $(-12)(3x+3)=$

17) $(5-3x)6=$

18) $4(2+6x)=$

19) $8(7x-3)=$

20) $(-2x+3)4=$

21) $(7-5x)(-9)=$

22) $(-10)(x-8)=$

23) $(11-4x)3=$

24) $(-6)(10x-4)=$

25) $(3-9x)(-7)=$

26) $(-9)(x+9)=$

27) $(-3+5x)(-7)=$

28) $(-5)(8-10x)=$

29) $12(4x-8)=$

30) $(-10x+13)(-3)=$

31) $(-8)(3x-2)+4(x+5)=$

32) $(-8)(x+4)-(6+5x)=$

Evaluating One Variable Expressions

Evaluate each expression using the value given.

1) $8 - x, x = 5$

2) $x - 9, x = 5$

3) $5x + 4, x = 3$

4) $x - 13, x = -4$

5) $12 - x, x = 4$

6) $x + 2, x = 6$

7) $4x + 8, x = 3$

8) $x + (-7), x = -8$

9) $4x + 5, x = 2$

10) $3x + 9, x = -2$

11) $15 + 3x - 7, x = 2$

12) $17 - 3x, x = 3$

13) $8x - 9, x = 4$

14) $5x + 4, x = -3$

15) $10x + 5, x = 3$

16) $14 - 4x, x = -6$

17) $3(5x + 3), x = 9$

18) $4(-3x - 6), x = 3$

19) $7x - 2x + 12, x = 4$

20) $(5x + 6) \div 2, x = 8$

21) $(x + 18) \div 10, x = 12$

22) $5x - 12 + 3x, x = -3$

23) $(6 - 4x)(-3), x = -4$

24) $9x^2 + 3x - 6, x = 2$

25) $x^2 - 10x, x = -5$

26) $3x(7 - 2x), x = 2$

27) $12x + 6 - 2x^2, x = -4$

28) $(-3)(4x - 8 + 3x), x = 3$

29) $(-6) + \frac{x}{4} + 3x, x = 16$

30) $(-6) + \frac{x}{5}, x = 35$

31) $\left(-\frac{45}{x}\right) - 7 + 2x, x = 9$

32) $\left(-\frac{21}{x}\right) - 12 + 4x, x = 7$

Evaluating Two Variables Expressions

Evaluate each expression using the values given.

1) $2x - 4y$,

$x = 4, y = 1$

2) $3x + 5y$,

$x = -2, y = 2$

3) $-7a + 4b$,

$a = 2, b = 4$

4) $3x + 5 - y$,

$x = 5, y = 6$

5) $3z + 12 - 2k$,

$z = 5, k = 6$

6) $6(-x - 3y)$,

$x = 5, y = -2$

7) $5a + 3b$,

$a = 3, b = 4$

8) $7x \div 3y$,

$x = 3, y = 7$

9) $2x + 15 + 5y$,

$x = -3, y = 1$

10) $5a - (18 - b)$,

$a = 2, b = 8$

11) $2z + 20 + 5k$,

$z = -6, k = 5$

12) $xy + 10 + 4x$,

$x = 3, y = 5$

13) $2x + 4y - 8 + 5$,

$x = 5, y = 2$

14) $\left(-\frac{24}{x}\right) + 3 + 2y$,

$x = 4, y = 6$

15) $(-3)(-3a - 3b)$,

$a = 4, b = 5$

16) $12 + 4x - 7 - y$,

$x = 3, y = 5$

17) $11x + 5 - 8y + 6$,

$x = 5, y = 2$

18) $10 + 2(-4x - 5y)$,

$x = 5, y = 4$

19) $5x + 13 + 6y$,

$x = 5, y = 6$

20) $10a - (7a + 3b) - 11$,

$a = 3, b = 8$

Combining like Terms

Simplify each expression.

1) $11x + 3x + 6 =$

2) $8(2x - 6) =$

3) $18x - 7x + 11 =$

4) $(-4)(6x - 7) =$

5) $22x - 10x - 5 =$

6) $32x - 13 + 8x =$

7) $15 - (8x - 11) =$

8) $-24x + 17 - 11x =$

9) $12x - 8 - 6x + 9 =$

10) $21x + 5 - 36 + 12x =$

11) $28x + 3x - 11 =$

12) $(-3x + 4)5 =$

13) $2 + 4x + 9x - 8 =$

14) $6(2x - 5x) - 4 =$

15) $4(5x + 11) + 3x =$

16) $x - 14 - 11x =$

17) $5(10 + 9x) - 8x =$

18) $42x + 17 - 23x =$

19) $(-7x) + 19 + 20x =$

20) $(-7x) - 33 + 29x =$

21) $4(5x + 3) - 19x =$

22) $5(6 - 2x) - 15x =$

23) $-24x + (11 - 18x) =$

24) $(-9) - (6)(7x + 3) =$

25) $(-1)(8x - 10) - 21x =$

26) $-36x + 14 + 27x - 5x =$

27) $3(-13x + 6) - 17x =$

28) $-5x - 42 + 32x =$

29) $37x - 19x + 15 - 9x =$

30) $3(5x + 7x) - 31 =$

31) $14 - 6x - 15 - 9x =$

32) $-2(-5x - 7x) + 27x =$

Answers of Worksheets

Simplifying Variable Expressions

1) $3x+15$
2) $-28x+20$
3) $5x+5$
4) $-8x^2-4$
5) $13x^2+10$
6) $18x^2+7x$
7) $-9x^2+4x$
8) $4x^2-10x$
9) $-22x+21$
10) $68x-12$
11) $-18x-71$
12) $-11x^2+5x$
13) $-6x+12$
14) $x-4$
15) $22x-5$
16) $14x-22$
17) $27x-16$
18) $25x+28$
19) $-23x-20$
20) $-3x^2-14x$
21) $-20x^2+27x$
22) $-20x^2-15x$
23) $4x^2+25x-19$
24) $6x^2-66x+25$
25) $3x^2+19x-5$
26) $-7x^2-20x$
27) $-12x^2+10x-7$
28) $-6x^2-21x+13$
29) $10x^2+7x+17$
30) $25x^2+25x$
31) $-16x^2-23x+29$
32) $-9x^2+3x+30$

Simplifying Polynomial Expressions

1) $2x^3+5x^2-11x$
2) $2x^5+2x^3-11x^2$
3) $21x^4+x^2$
4) $12x^3+8x^2+13x$
5) $-26x^3+15x^2-22$
6) $-5x^4-6x^3-3x$
7) $-26x^3-10x^2+25x$
8) $-5x^3-11x^2-10x$
9) $-3x^4+5x^3+2x^2-5x$
10) $11x^4-3x^2+2x$
11) $-13x^4+3x^2$
12) $19x^4-29x^3+2x^2$
13) $3x^4-32x^3+10x^2$
14) $17x^3-13x^2-6x$
15) $-11x^5-29x^4+5x^2$
16) $10x^3-5x^2+13x$

Translate Phrases into an Algebraic Statement

1) $9x$
2) $y-11$
3) $\frac{19}{x}$
4) $38-y$
5) $y+40$
6) 6^2
7) x^5
8) $6+x$
9) $57-y$
10) $\frac{9}{x}$
11) $\frac{x^2}{25}$
12) $x-6=19$
13) $10a-b^2$
14) $41-ab$

The Distributive Property

1) $8x+4$
2) $14x+8$
3) $12x-12$
4) $-12x+30$
5) $-3x-18$
6) $6x+8$
7) $15x-40$
8) $7x+5$
9) $18x-9$
10) $-4x+28$
11) $3x-5$
12) $12x+27$

13) $18x + 24$
14) $-10x + 6$
15) $24x - 15$
16) $-36x - 36$
17) $-18x + 30$
18) $24x + 8$
19) $56x - 24$
20) $-8x + 12$
21) $45x - 63$
22) $-10x + 80$
23) $-12x + 33$
24) $-60x + 24$
25) $63x - 21$
26) $-9x - 81$
27) $-35x + 21$
28) $50x - 40$
29) $48x - 96$
30) $30x - 39$
31) $-20x + 36$
32) $-13x - 38$

Evaluating One Variables

1) 3
2) -4
3) 19
4) -17
5) 8
6) 8
7) 20
8) -15
9) 13
10) 3
11) 14
12) 8
13) 23
14) -11
15) 35
16) 38
17) 144
18) -60
19) 32
20) 23
21) 3
22) -36
23) -66
24) 36
25) 75
26) 18
27) -74
28) -39
29) 46
30) 1
31) 6
32) 13

Evaluating Two Variables

1) 4
2) 4
3) 2
4) 14
5) 15
6) 6
7) 27
8) 1
9) 14
10) 0
11) 33
12) 37
13) 15
14) 9
15) 81
16) 12
17) 50
18) -70
19) 74
20) -26

Combining like Terms

1) $14x + 6$
2) $16x - 48$
3) $11x + 11$
4) $-24x + 28$
5) $12x - 5$
6) $40x - 13$
7) $-8x + 26$
8) $-35x + 17$
9) $6x + 1$
10) $33x - 31$
11) $31x - 11$
12) $-15x + 20$
13) $13x - 6$
14) $-18x - 4$
15) $23x + 44$
16) $-10x - 14$
17) $37x + 50$
18) $19x + 17$
19) $13x + 19$
20) $22x - 33$
21) $x + 12$
22) $-25x + 30$
23) $-42x + 11$
24) $-42x - 27$
25) $-29x + 10$
26) $-14x + 14$
27) $-56x + 18$
28) $27x - 42$
29) $9x + 15$
30) $36x - 31$
31) $-15x - 1$
32) $51x$

Chapter 6

Equations and Inequalities

Topics that you will practice in this chapter:

- ✓ One–Step Equations
- ✓ Multi–Step Equations
- ✓ Graphing Single–Variable Inequalities
- ✓ One–Step Inequalities
- ✓ Multi-Step Inequalities
- ✓ Systems of Equations
- ✓ Systems of Equations Word Problems

"Life is a math equation. In order to gain the most, you have to know how to convert negatives into positives." – Anonymous

One–Step Equations

Find the answer for each equation.

1) $3x = 90, x =$ ____

2) $5x = 35, x =$ ____

3) $6x = 24, x =$ ____

4) $24x = 144, x =$ ____

5) $x + 15 = 20, x =$ ____

6) $x - 7 = 4, x =$ ____

7) $x - 9 = 2, x =$ ____

8) $x + 15 = 23, x =$ ____

9) $x - 4 = 13, x =$ ____

10) $12 = 16 + x, x =$ ____

11) $x - 10 = 2, x =$ ____

12) $5 - x = -11, x =$ ____

13) $28 = -6 + x, x =$ ____

14) $x - 20 = -35, x =$ ____

15) $x + 14 = -4, x =$ ____

16) $14 = 28 - x, x =$ ____

17) $7 + x = -7, x =$ ____

18) $x - 16 = 4, x =$ ____

19) $30 = x - 15, x =$ ____

20) $x - 5 = -18, x =$ ____

21) $x - 10 = 24, x =$ ____

22) $x - 20 = -25, x =$ ____

23) $x - 17 = 30, x =$ ____

24) $-70 = x - 28, x =$ ____

25) $x - 9 = 13, x =$ ____

26) $36 = 4x, x =$ ____

27) $x - 35 = 25, x =$ ____

28) $x - 25 = 10, x =$ ____

29) $70 - x = 16, x =$ ____

30) $x - 10 = 14, x =$ ____

31) $17 - x = -13, x =$ __

32) $\mathrm{x} - 9 = -30, \mathrm{x} =$ ____

Multi–Step Equations

Find the answer for each equation.

1) $3x + 3 = 9$

2) $-x\ + 5 = 12$

3) $4x - 8 = 8$

4) $-(3 - x) = 5$

5) $4x - 8 = 16$

6) $12x - 15 = 9$

7) $2x - 18 = 2$

8) $4x + 8 = 16$

9) $24x + 27 = 75$

10) $-14(3 + x) = 14$

11) $-3(2 + x) = 6$

12) $12 = -\ (x - 7)$

13) $3(3 - x) = 30$

14) $-15 = -(3x + 6)$

15) $40(3 + x) = 40$

16) $5(x - 10) = 25$

17) $-18 = x + 8x$

18) $3x + 25 = -2x - 10$

19) $7(6 + 3x) = -63$

20) $18 - 3x = -4 - 5x$

21) $4 - 6x = 36 + 2x$

22) $15 + 15x = -5 + 5x$

23) $42 = (-6x) - 7 + 7$

24) $21 = 3x - 21 + 4x$

25) $-18 = -6x - 9 + 3x$

26) $5x - 15 = -29 + 6x$

27) $7x - 18 = 4x + 3$

28) $-7 - 4x = 5(4 - x)$

29) $x - 5 = -5(-3 - x)$

30) $13x - 68 = 15x - 102$

31) $-5x - 3 = -3(9 + 3x)$

32) $-2\text{x} - 15 = 6\text{x} + 17$

Graphing Single–Variable Inequalities

Draw a graph for each inequality.

1) $x > -1$

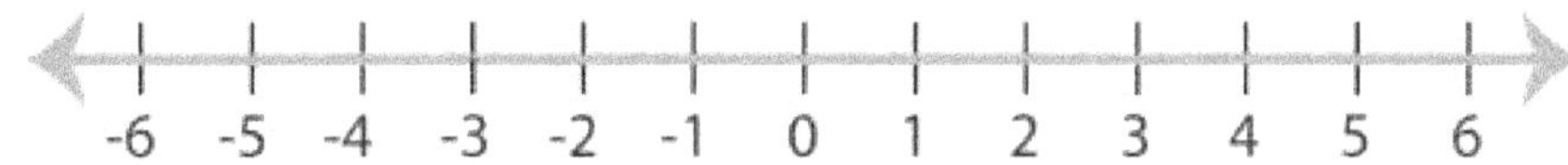

2) $x \leq 2$

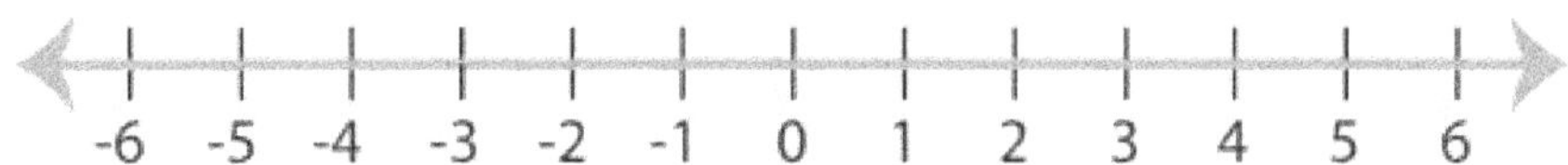

3) $x \geq 0$

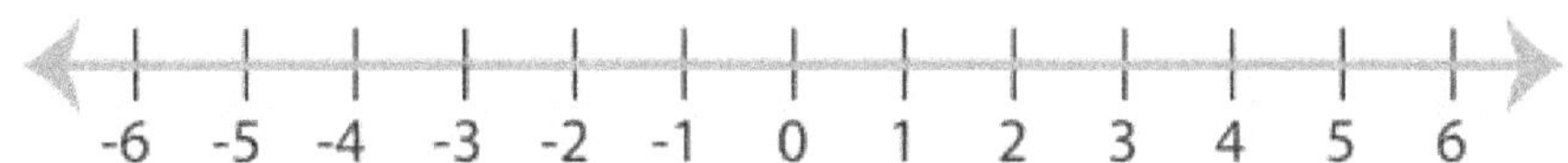

4) $x < -3$

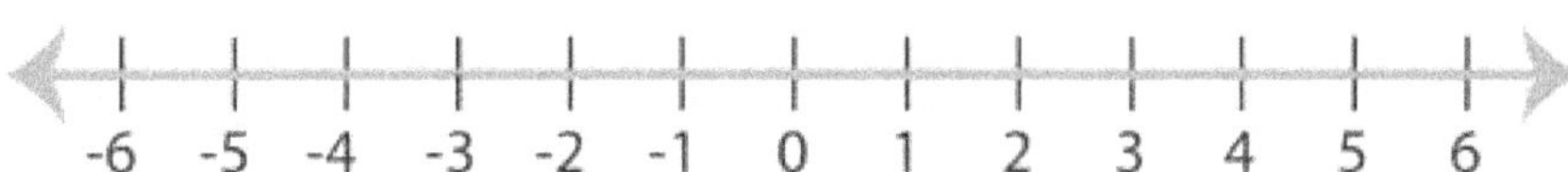

5) $x < \frac{1}{2}$

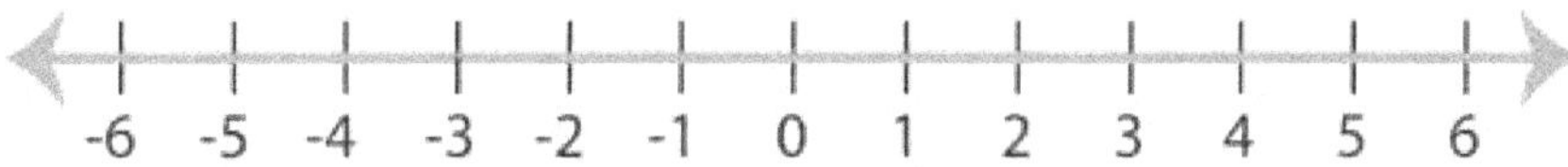

6) $x \leq -2$

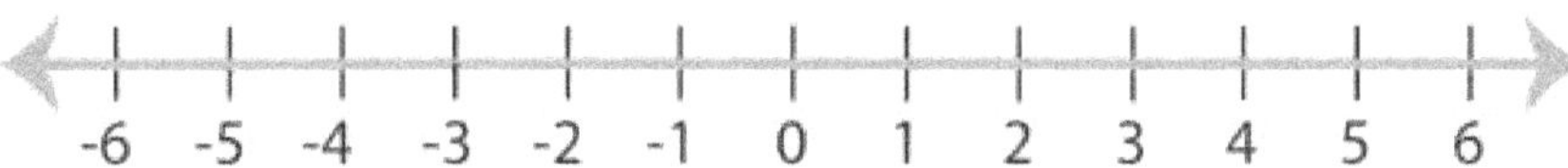

7) $x \leq 3$

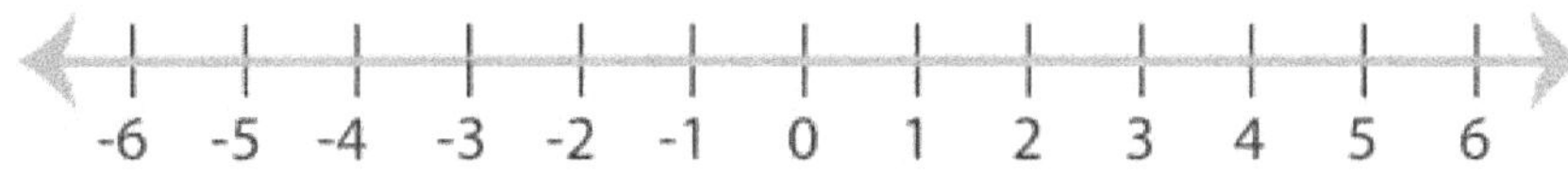

8) $x \geq -\frac{7}{2}$

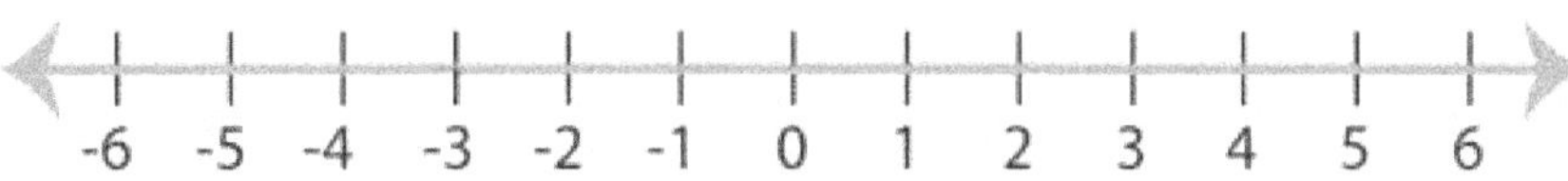

One–Step Inequalities

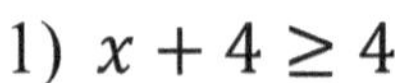 Find the answer for each inequality and graph it.

1) $x + 4 \geq 4$

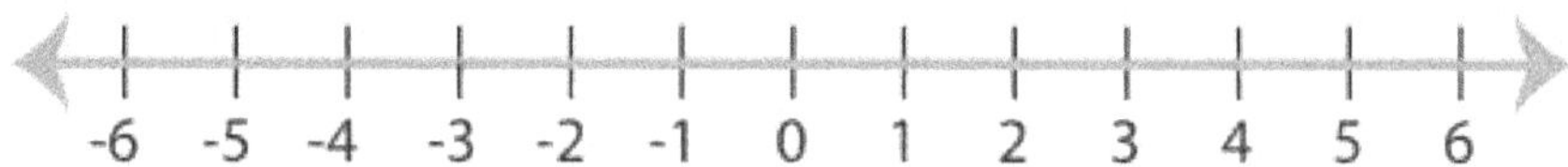

2) $x - 5 \leq 2$

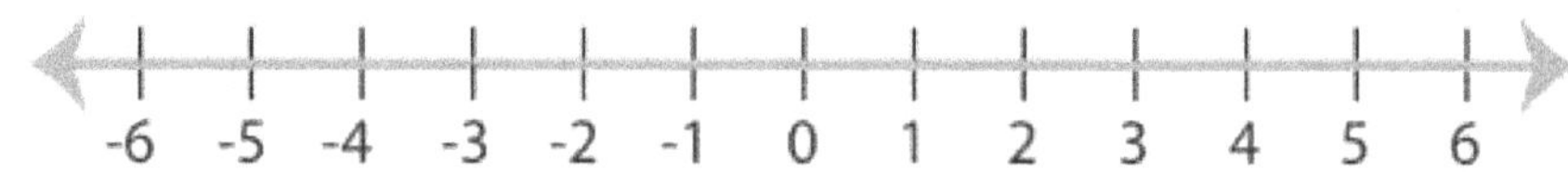

3) $5x > 35$

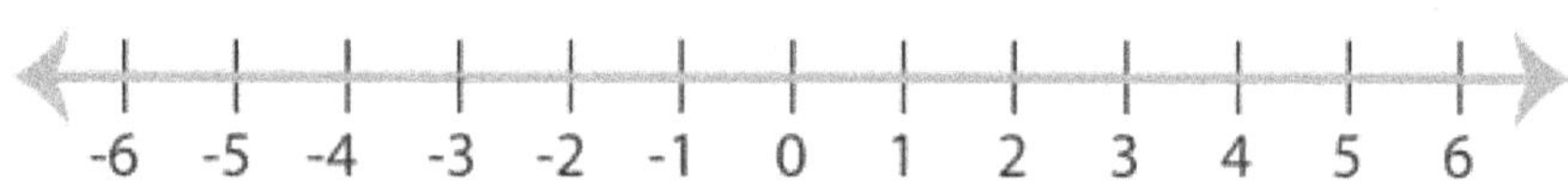

4) $9 + x \leq 11$

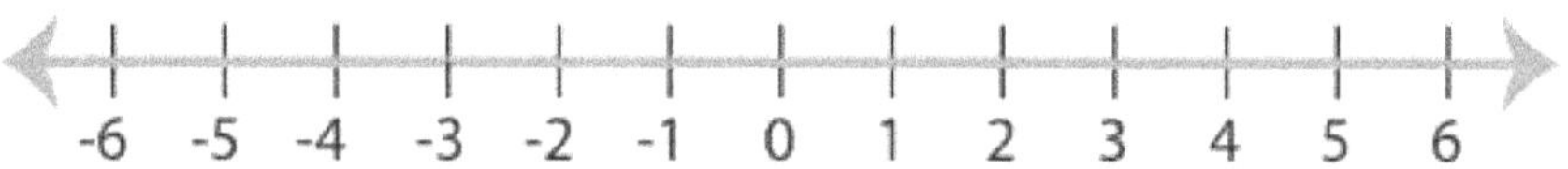

5) $x - 5 < -9$

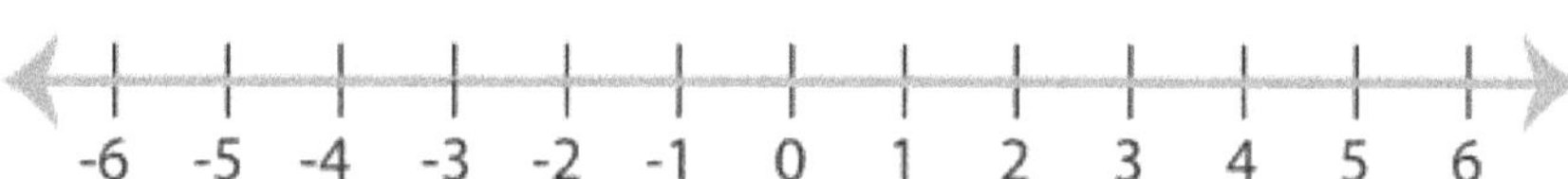

6) $9x \geq 72$

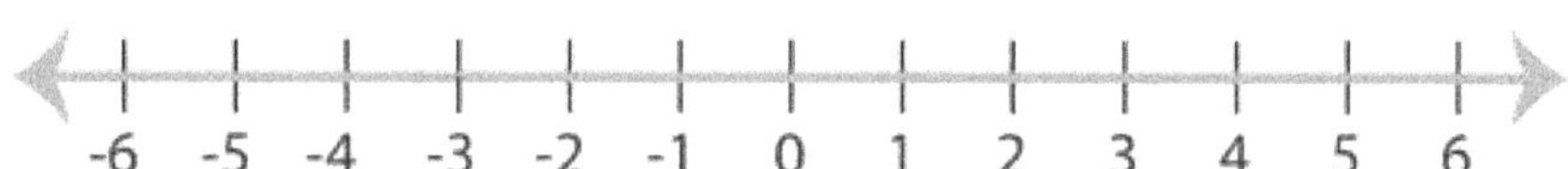

7) $9x \leq 27$

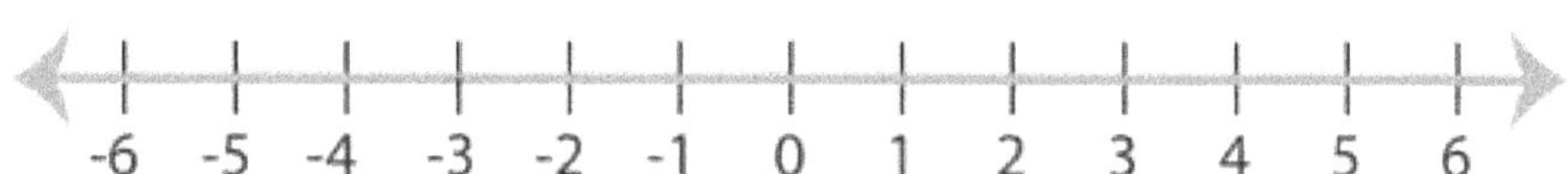

8) $x + 19 > 16$

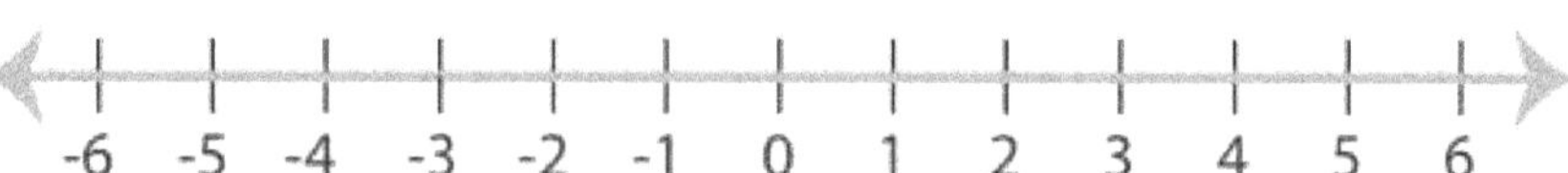

Multi-Step Inequalities

Calculate each inequality.

1) $x - 3 \leq 7$

2) $8 - x \leq 8$

3) $3x - 9 \leq 9$

4) $4x - 4 \geq 8$

5) $x - 7 \geq 1$

6) $5x - 15 \leq 5$

7) $6x - 8 \leq 4$

8) $-11 + 6x \leq 12$

9) $4(x - 4) \leq 16$

10) $3x - 10 \leq 11$

11) $5x - 25 < 25$

12) $9x - 5 < 22$

13) $20 - 7x \geq -15$

14) $33 + 6x < 45$

15) $8 + 8x \geq 96$

16) $7 + 3x < 13$

17) $4x - 3 < 9$

18) $5(2 - 2x) \geq -30$

19) $-(7 + 6x) < 29$

20) $12 - 8x \geq -20$

21) $-4(x - 6) > 24$

22) $\frac{3x + 9}{6} \leq 10$

23) $\frac{4x - 10}{3} \leq 2$

24) $\frac{2x - 8}{3} > 2$

25) $8 + \frac{x}{6} < 9$

26) $\frac{9x}{7} - 4 < 5$

27) $\frac{15x + 45}{15} > 1$

28) $16 + \frac{x}{4} < 6$

Systems of Equations

Calculate each system of equations.

1) $-x + y = 2$ $\quad x =$ ____
$-4x + 2y = 6$ $\quad y =$ ____

2) $-15x + 3y = -9$ $\quad x =$ ____
$9x - 16y = 48$ $\quad y =$ ____

3) $y = -7$ $\quad x =$ ____
$6x + 5y = 7$ $\quad y =$ ____

4) $3y = -9x + 15$ $\quad x =$ ____
$5x - 4y = -3$ $\quad y =$ ____

5) $10x - 9y = -13$ $\quad x =$ ____
$-5x + 3y = 11$ $\quad y =$ ____

6) $-12x - 16y = 20$ $\quad x =$ ____
$6x - 12y = 30$ $\quad y =$ ____

7) $5x - 14y = -23$ $\quad x =$ ____
$-18x + 21y = 24$ $\quad y =$ ____

8) $15x - 21y = -6$ $\quad x =$ ____
$2x - 3y = -2$ $\quad y =$ ____

9) $-x + 3y = 3$ $\quad x =$ ____
$-14x + 16y = -10$ $\quad y =$ ____

10) $x + 5y = 50$ $\quad x =$ ____
$3x + 10y = 80$ $\quad y =$ ____

11) $6x - 7y = -8$ $\quad x =$ ____
$-x - 4y = -9$ $\quad y =$ ____

12) $2x + 4y = -10$ $\quad x =$ ____
$2x - 8y = 14$ $\quad y =$ ____

13) $4x + 3y = 12$ $\quad x =$ ____
$5x - 3y = 15$ $\quad y =$ ____

14) $3x - 2y = 3$ $\quad x =$ ____
$7x - 8y = 22$ $\quad y =$ ____

15) $3x + 2y = 5$ $\quad x =$ ____
$-10x - 4y = -14$ $\quad y =$ ____

16) $10x + 7y = 1$ $\quad x =$ ____
$-5x - 7y = 24$ $\quad y =$ ____

Systems of Equations Word Problems

Find the answer for each word problem.

1) Tickets to a movie cost $4 for adults and $3 for students. A group of friends purchased 8 tickets for $31.00. How many adults ticket did they buy? ____
2) At a store, Eva bought two shirts and five hats for $77.00. Nicole bought three same shirts and four same hats for $84.00. What is the price of each shirt? _____
3) A farmhouse shelters 18 animals, some are pigs, and some are ducks. Altogether there are 66 legs. How many pigs are there? _____
4) A class of 214 students went on a field trip. They took 36 vehicles, some cars and some buses. If each car holds 5 students and each bus hold 22 students, how many buses did they take? _____
5) A theater is selling tickets for a performance. Mr. Smith purchased 5 senior tickets and 3 child tickets for $105 for his friends and family. Mr. Jackson purchased 3 senior tickets and 5 child tickets for $79. What is the price of a senior ticket? $_____
6) The difference of two numbers is 10. Their sum is 20. What is the bigger number? $_____
7) The sum of the digits of a certain two–digit number is 7. Reversing its digits increase the number by 9. What is the number? _____
8) The difference of two numbers is 11. Their sum is 25. What are the numbers? __________
9) The length of a rectangle is 5 meters greater than 2 times the width. The perimeter of rectangle is 28 meters. What is the length of the rectangle? __________
10) Jim has 25 nickels and dimes totaling $1.80. How many nickels does he have? _____

Answers of Worksheets

One–Step Equations

1) 30	9) 17	17) −14	25) 22
2) 7	10) −4	18) 20	26) 9
3) 4	11) 12	19) 45	27) 60
4) 6	12) 16	20) −13	28) 35
5) 5	13) 34	21) 34	29) 54
6) 11	14) −15	22) −5	30) 24
7) 11	15) −18	23) 47	31) 30
8) 8	16) 14	24) −42	32) −21

Multi–Step Equations

1) 2	9) 2	17) −2	25) 3
2) −7	10) −4	18) −7	26) 14
3) 4	11) −4	19) −5	27) 7
4) 8	12) −5	20) −11	28) 27
5) 6	13) −7	21) −4	29) −5
6) 2	14) 3	22) −2	30) 17
7) 10	15) −2	23) −7	31) −6
8) 2	16) 15	24) 6	32) −4

Graphing Single–Variable Inequalities

1)

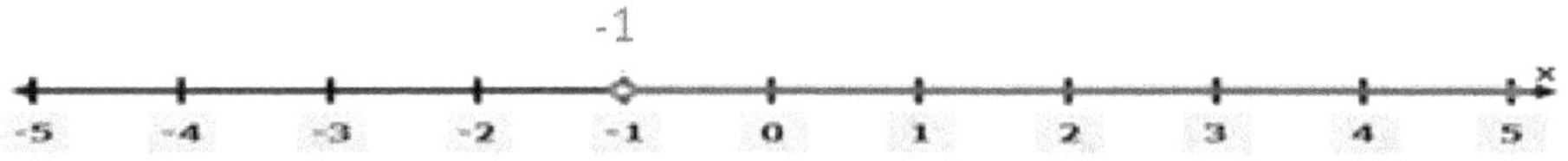

-1

-5 -4 -3 -2 -1 0 1 2 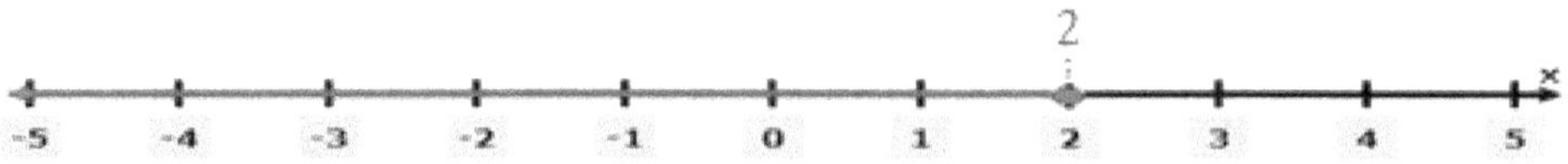 3 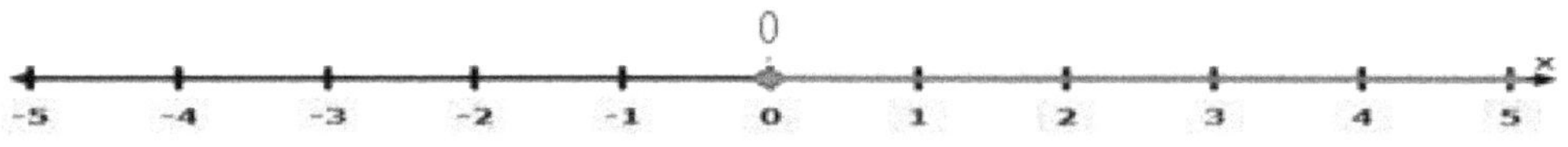 4 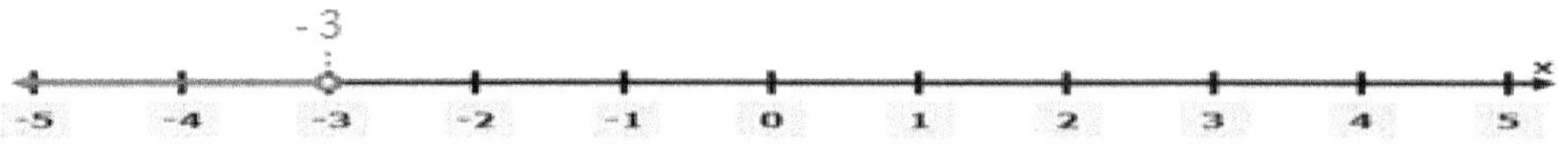 5 x

2)

2

-5 -4 -3 -2 -1 0 1 2 3 4 5 x

3)

0

-5 -4 -3 -2 -1 0 1 2 3 4 5 x

4)

-3

-5 -4 -3 -2 -1 0 1 2 3 4 5 x

5)

6)

7)

8)

One–Step Inequalities

1) 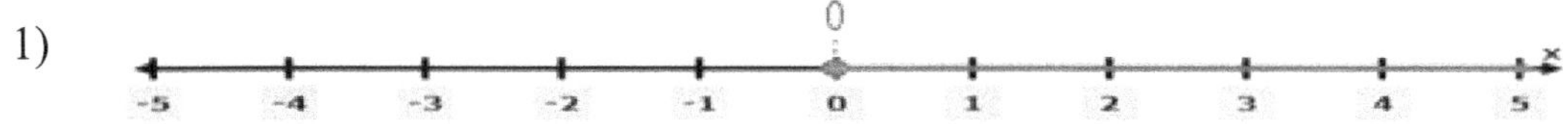

2)

3)

4)

5)

6)

7)

8)

Multi-Step Inequalities

1) $x \le 10$
2) $x \ge 0$
3) $x \le 6$
4) $x \ge 3$
5) $x \ge 8$
6) $x \le 4$
7) $x \le 2$
8) $x \le \frac{23}{6}$
9) $x \le 8$
10) $x \le 7$
11) $x < 10$
12) $x < 3$
13) $x \le 5$
14) $x < 2$
15) $x \ge 11$
16) $x < 2$
17) $x < 3$
18) $x \le 4$
19) $x > -6$
20) $x \le 4$
21) $x < 0$
22) $x \le 17$
23) $x \le 4$
24) $x > 7$

25) $x < 6$ 26) $x < 7$ 27) $x > -2$ 28) $x < -40$

Systems of Equations

1) $x = -1, y = 1$
2) $x = 0, y = -3$
3) $x = 7$
4) $x = 1, y = 2$
5) $x = -4, y = -3$
6) $x = 1, y = -2$
7) $x = 1, y = 2$
8) $x = 8, y = 6$
9) $x = 3, y = 2$
10) $x = -20, y = 14$
11) $x = 1, y = 2$
12) $x = -1, y = -2$
13) $x = 3, y = 0$
14) $x = -2, y = -\frac{9}{2}$
15) $x = 1, y = 1$
16) $x = 5, y = -7$

Systems of Equations Word Problems

1) 7
2) $16
3) 15
4) 2
5) $18
6) 15
7) 34
8) 18, 7
9) 11 meters
10) 14

Chapter 7 :

Linear Functions

Topics that you will practice in this chapter:

- ✓ Finding Slope
- ✓ Graphing Lines Using Line Equation
- ✓ Writing Linear Equations
- ✓ Graphing Linear Inequalities
- ✓ Finding Midpoint
- ✓ Finding Distance of Two Points

"Nature is written in mathematical language." – Galileo Galilei

Finding Slope

Find the slope of each line.

1) $y = x + 8$

2) $y = -3x + 5$

3) $y = 2x + 12$

4) $y = -4x + 19$

5) $y = 11 + 6x$

6) $y = 7 - 5x$

7) $y = 8x + 19$

8) $y = -9x + 20$

9) $y = -7x + 4$

10) $y = 3x - 8$

11) $y = \frac{1}{3}x + 8$

12) $y = -\frac{4}{5}x + 9$

13) $-3x + 6y = 30$

14) $4x + 4y = 16$

15) $3y - x = 10$

16) $8y - x = 5$

Find the slope of the line through each pair of points.

17) $(2, 3), (7, 10)$

18) $(-3,5), (2, 15)$

19) $(5, -3), (1,9)$

20) $(-5, -5), (10, 25)$

21) $(22, 3), (7, 18)$

22) $(-16, 8), (-7, 26)$

23) $(25, 11), (29, 19)$

24) $(26, -19), (14, 17)$

25) $(22, -13), (20, -11)$

26) $(19, 7), (15, -3)$

27) $(5, 7), (11, 19)$

28) $(52, -62), (40, 70)$

Graphing Lines Using Line Equation

Sketch the graph of each line.

1) $y = x - 2$

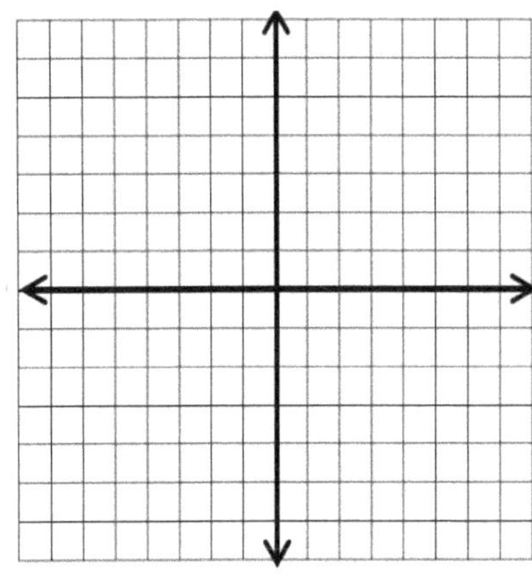

2) $y = -3x + 2$

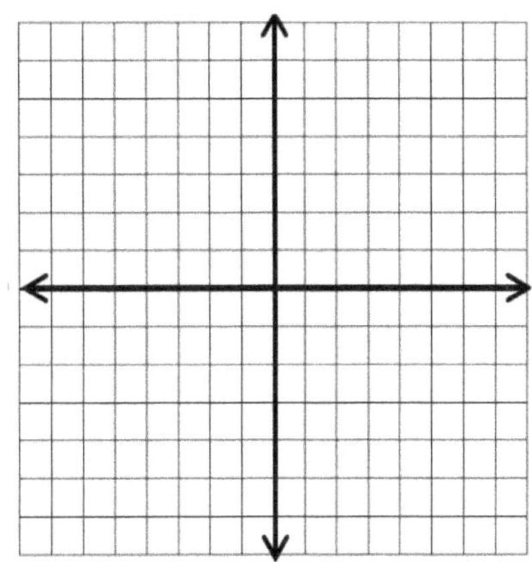

3) $x + y = 0$

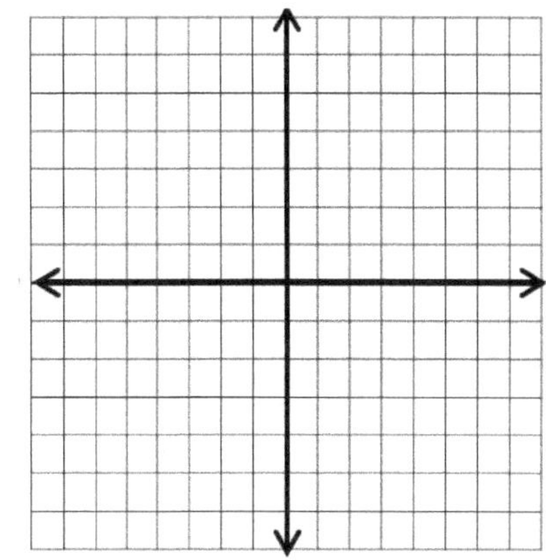

4) $x + y = -3$

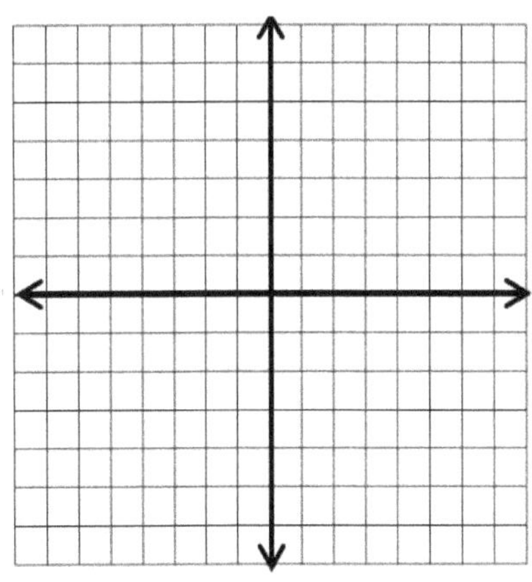

5) $2x + 3y = -4$

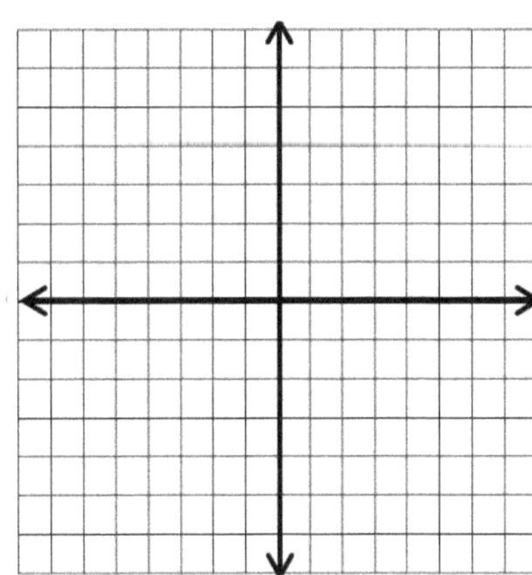

6) $y - 3x + 6 = 0$

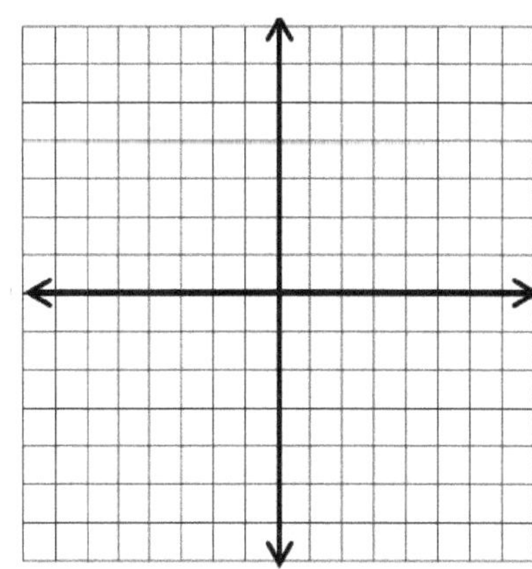

Writing Linear Equations

Write the equation of the line through the given points.

1) Through: $(2, -5), (3, 9)$

2) Through: $(-6, 3), (3, 12)$

3) Through: $(10, 7), (5, 27)$

4) Through: $(15, 11), (3, -1)$

5) Through: $(24, 17), (12, -7)$

6) Through: $(8, 29), (4, -7)$

7) Through: $(20, -16), (12, 0)$

8) Through: $(-3, 10), (2, -5)$

9) Through: $(-6, 17), (4, -3)$

10) Through: $(-8, 22), (5, -4)$

11) Through: $(9, 27), (3, -3)$

12) Through: $(11, 32), (9, 4)$

13) Through: $(-3, 13), (-4, 0)$

14) Through: $(-5, 5), (5, 15)$

15) Through: $(18, -32), (11, 3)$

16) Through: $(-4, 25), (4, -15)$

Find the answer for each problem.

17) What is the equation of a line with slope 6 and intercept 12?

____

18) What is the equation of a line with slope -11 and intercept -4?

____

19) What is the equation of a line with slope -3 and passes through point $(5, 2)$? ______________

20) What is the equation of a line with slope -5 and passes through point $(-2, -1)$? ______________

21) The slope of a line is -10 and it passes through point $(-3, 0)$. What is the equation of the line? ______________

22) The slope of a line is 8 and it passes through point $(0, 7)$. What is the equation of the line? ______________

Graphing Linear Inequalities

Sketch the graph of each linear inequality.

1) $y > 4x - 5$ 2) $y < 2x + 4$ 3) $y \leq -5x - 2$

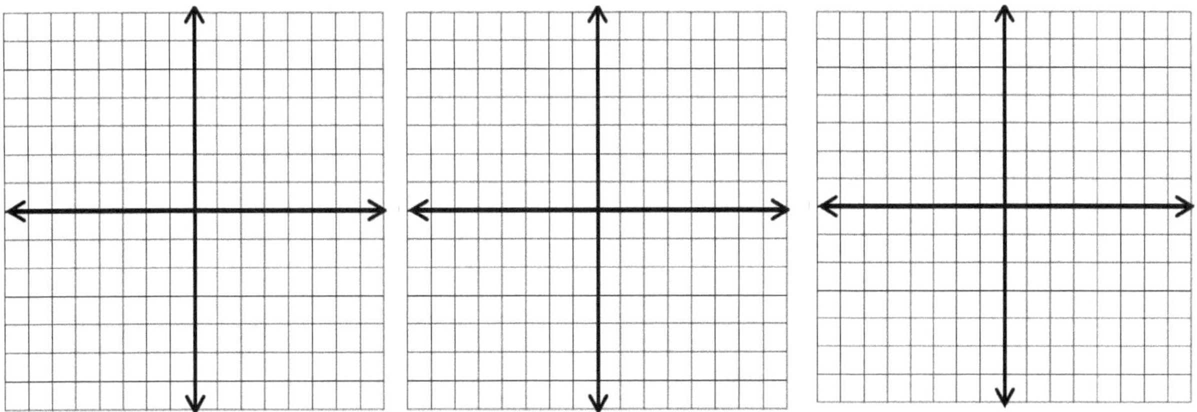

4) $4y \geq 12 + 4x$ 5) $-12y < 3x - 24$ 6) $5y \geq -15x + 10$

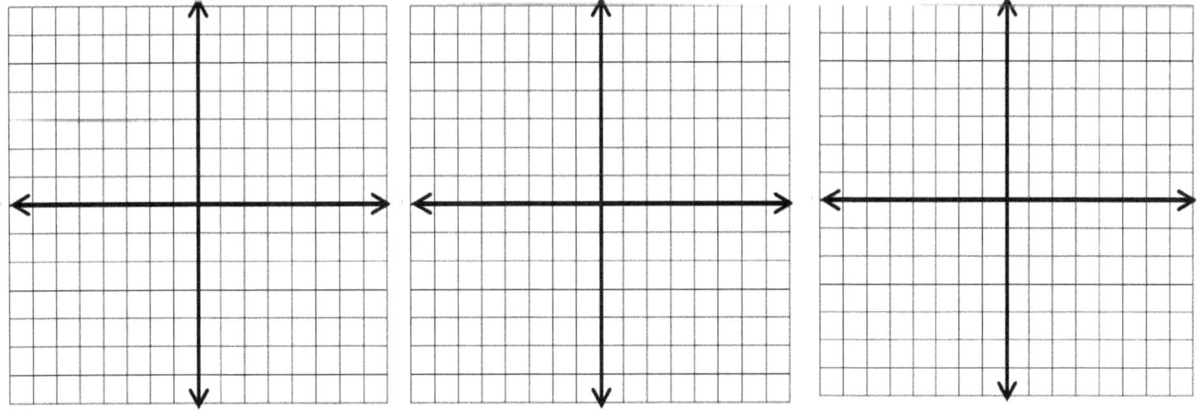

Finding Midpoint

Find the midpoint of the line segment with the given endpoints.

1) $(-4,-3),(2,3)$

2) $(9,0),(-1,8)$

3) $(9,-6),(3,14)$

4) $(-10,-6),(0,8)$

5) $(2,-5),(14,-15)$

6) $(-10,-3),(4,-13)$

7) $(8,7),(-8,13)$

8) $(-3,6),(-9,2)$

9) $(-4,5),(16,-9)$

10) $(7,14),(9,-2)$

11) $(-8,6),(6,6)$

12) $(10,5),(-2,-3)$

13) $(-5,12),(-3,3)$

14) $(12,7),(8,-2)$

15) $(10,2),(-6,14)$

16) $(-1,-2),(-7,10)$

17) $(7,-7),(13,-13)$

18) $(-3,-8),(11,-4)$

19) $(5,-11),(-8,9)$

20) $(14,-4),(16,14)$

21) $(0,-5),(8,-1)$

22) $(3,0),(-21,18)$

23) $(17,-3),(-7,-5)$

24) $(26,-12),(6,24)$

Find the answer for each problem.

25) One endpoint of a line segment is $(-3,7)$ and the midpoint of the line segment is $(-6,9)$. What is the other endpoint? ______________

26) One endpoint of a line segment is $(-3,7)$ and the midpoint of the line segment is $(1,5)$. What is the other endpoint? ______________

27) One endpoint of a line segment is $(-10,-16)$ and the midpoint of the line segment is $(2,9)$. What is the other endpoint? ______________

Finding Distance of Two Points

Find the distance between each pair of points.

1) $(6, 3), (-3, -9)$

2) $(5, 2), (-10, -6)$

3) $(8, 5), (8, 3)$

4) $(-8, -2), (2, 22)$

5) $(6, -7), (-3, -7)$

6) $(12, 0), (-9, -20)$

7) $(3, 20), (3, -5)$

8) $(10, 17), (5, 5)$

9) $(7, -2), (-4, -2)$

10) $(13, 4), (5, -2)$

11) $(11, 13), (5, 5)$

12) $(1, 4), (-23, -3)$

13) $(9,8), (5, -4)$

14) $(-11, -4), (5, 8)$

15) $(-2, -6), (-2, -12)$

16) $(-1, -4), (23, 3)$

17) $(19, 3), (7, -6)$

18) $(-5, -2), (3, 4)$

19) $(2, 6), (2, -12)$

20) $(-4, -2), (8, -2)$

Find the answer for each problem.

21) Triangle ABC is a right triangle on the coordinate system and its vertices are $(-2, 5)$, $(-2, 1)$, and $(1, 1)$. What is the area of triangle ABC? ______________

22) Three vertices of a triangle on a coordinate system are $(3, -6)$, $(-5, -12)$, and $(3, -18)$. What is the perimeter of the triangle? ________

23) Four vertices of a rectangle on a coordinate system are $(-2, 2)$, $(-2, 6)$, $(4, 2)$, and $(4, 6)$. What is its perimeter? ______________

Answers of Worksheets

Finding Slope

1) 1
2) -3
3) 2
4) -4
5) 6
6) -5
7) 8
8) -9
9) -7
10) 3
11) $\frac{1}{3}$
12) $-\frac{4}{5}$
13) $\frac{1}{2}$
14) -1
15) $\frac{1}{3}$
16) $\frac{1}{8}$
17) $\frac{7}{5}$
18) 2
19) -3
20) 2
21) -1
22) 2
23) 2
24) -3
25) -1
26) $\frac{5}{2}$
27) 2
28) -11

Graphing Lines Using Line Equation

1) $y = x - 2$

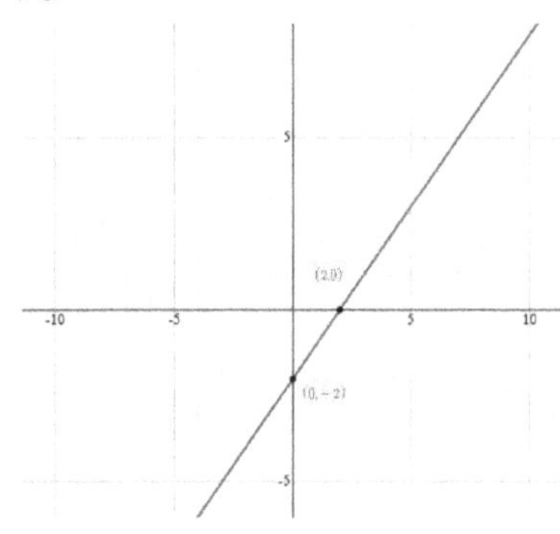

2) $y = -3x + 2$

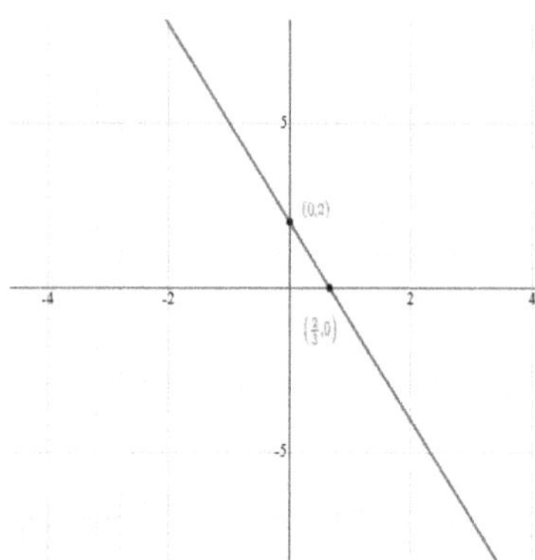

3) $x + y = 0$

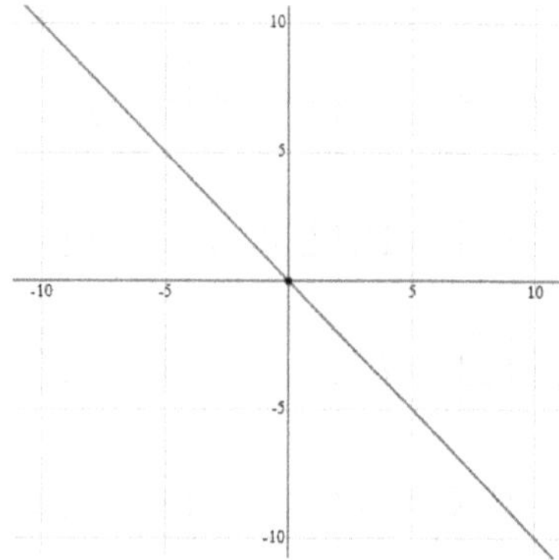

4) $x + y = -3$

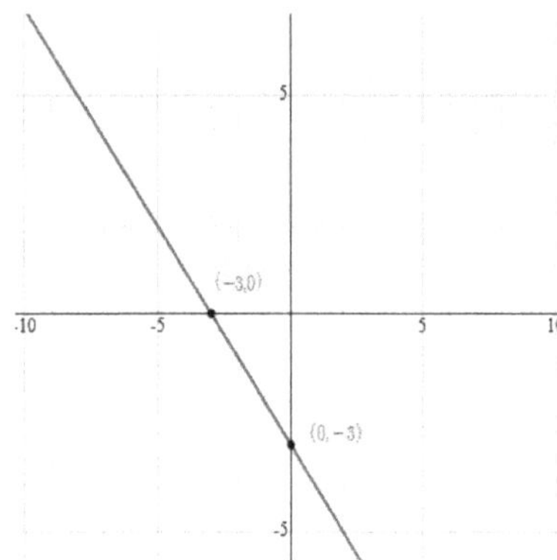

5) $2x + 3y = -4$

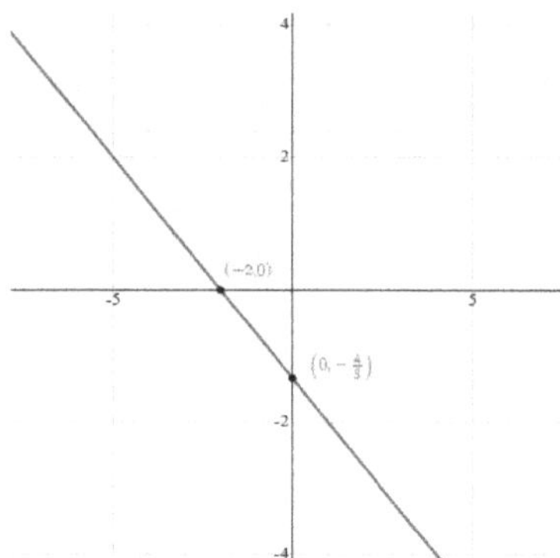

6) $y - 3x + 6 = 0$

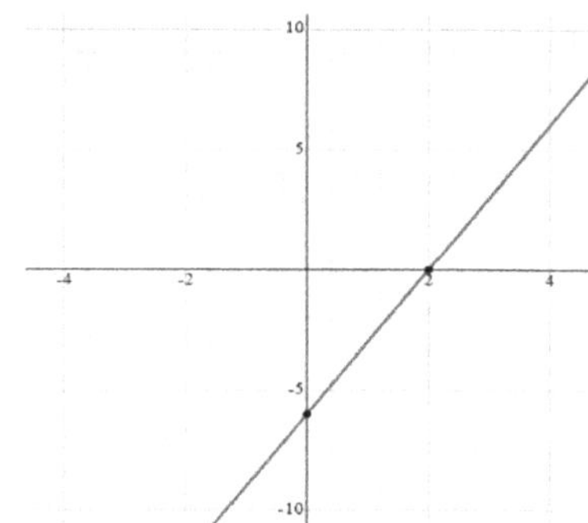

Writing Linear Equations

1) $y = 14x - 33$
2) $y = x + 9$
3) $y = -4x + 47$
4) $y = x - 4$
5) $y = 2x - 31$
6) $y = 9x - 43$
7) $y = -2x + 24$
8) $y = -3x + 1$
9) $y = -2x + 5$
10) $y = -2x + 6$
11) $y = 5x - 18$
12) $y = 14x - 122$
13) $y = 13x + 52$
14) $y = x + 10$
15) $y = -5x + 58$
16) $y = -5x + 5$
17) $y = 6x + 12$
18) $y = -11x - 4$
19) $y = -3x + 17$
20) $y = -5x - 11$
21) $y = -10x - 30$
22) $y = 8x + 7$

Graphing Linear Inequalities

1) $y > 4x - 5$

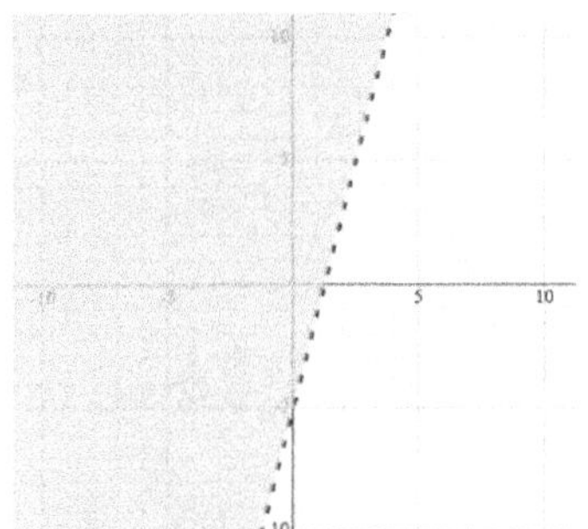

2) $y < 2x + 4$

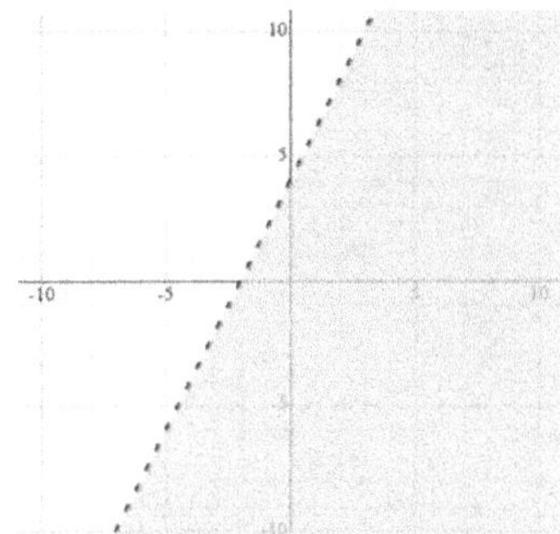

3) $y \leq -5x - 2$

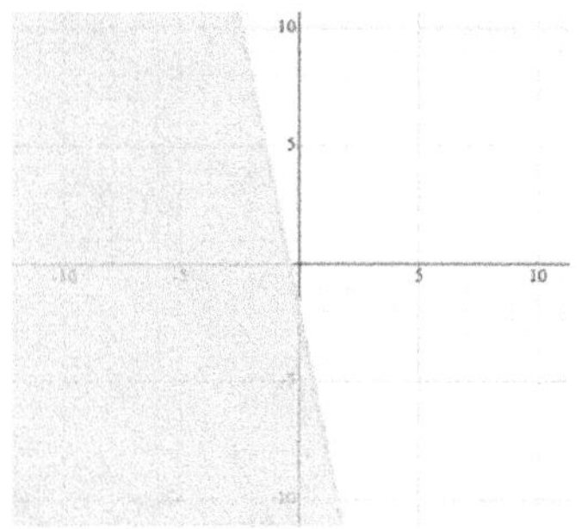

4) $4y \geq 12 + 4x$

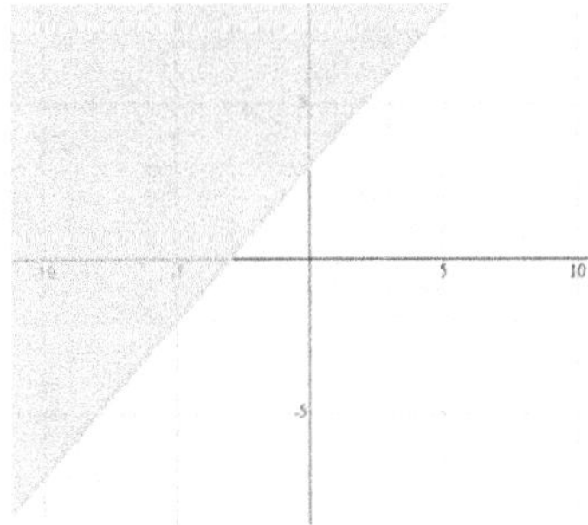

5) $-12y < 3x - 24$

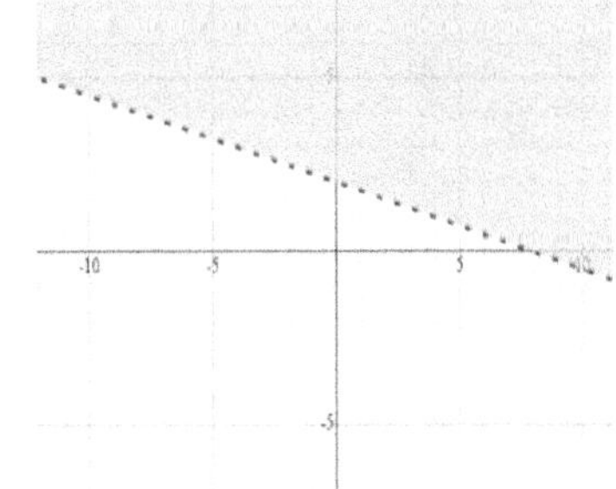

6) $5y \geq -15x + 10$

Finding Midpoint

1) $(-1, 0)$
2) $(4, 4)$
3) $(6, 4)$
4) $(-5, 1)$
5) $(8, -10)$
6) $(-3, -8)$
7) $(0, 10)$
8) $(-6, 4)$
9) $(6, -2)$
10) $(8, 6)$
11) $(-1, 6)$
12) $(4, 1)$
13) $(-4, 7.5)$
14) $(10, 2.5)$
15) $(2, 8)$
16) $(-4, 4)$
17) $(10, -10)$
18) $(4, -6)$

19) $(-1.5, -1)$
20) $(15, 5)$
21) $(4, -3)$
22) $(-9, 9)$
23) $(5, -4)$
24) $(16, 6)$
25) $(-9, 11)$
26) $(5, 3)$
27) $(14, 34)$

Finding Distance of Two Points

1) 15
2) 17
3) 2
4) 26
5) 9
6) 29
7) 25
8) 13
9) 11
10) 10
11) 10
12) 25
13) $4\sqrt{10}$
14) 20
15) 6
16) 25
17) 15
18) 10
19) 18
20) 12
21) 6 *square units*
22) 32 *units*
23) 20 *units*

Chapter 8 :
Polynomials

Topics that you will practice in this chapter:

- ✓ Writing Polynomials in Standard Form
- ✓ Simplifying Polynomials
- ✓ Adding and Subtracting Polynomials
- ✓ Multiplying Monomials
- ✓ Multiplying and Dividing Monomials
- ✓ Multiplying a Polynomial and a Monomial
- ✓ Multiplying Binomials
- ✓ Factoring Trinomials
- ✓ Operations with Polynomials

Mathematics is the supreme judge; from its decisions there is no appeal. – Tobias Dantzig

Writing Polynomials in Standard Form

Write each polynomial in standard form.

1) $11x - 7x =$

2) $-5 + 19x - 19x =$

3) $6x^5 - 12x^3 =$

4) $12 + 17x^4 - 12 =$

5) $5x^2 + 4x - 9x^3 =$

6) $-3x^2 + 12x^5 =$

7) $5x + 8x^3 - 2x^8 =$

8) $-7x^3 + 4x - 9x^6 =$

9) $3x^2 + 22 - 6x =$

10) $3 - 4x + 9x^4 =$

11) $13x^2 + 28x - 8x^3 =$

12) $16 + 4x^2 - 2x^3 =$

13) $19x^2 - 9x + 9x^4 =$

14) $3x^4 - 7x^2 - 2x^3 =$

15) $-51 + 3x^2 - 8x^4 =$

16) $7x^2 - 8x^6 + 4x^4 - 15 =$

17) $6x^4 - 4x^5 + 16 - 3x^3 =$

18) $-2x^6 + 4x - 7x^2 - 5x =$

19) $11x^7 + 8x^5 - 5x^7 - 3x^2 =$

20) $2x^2 - 12x^5 + 8x^2 + 3x^6 =$

21) $4x^5 - 11x^7 - 6x^3 + 16x^5 =$

22) $6x^3 + 3x^5 + 34x^4 - 8x^5 =$

23) $3x(4x + 5 - 2x^2) =$

24) $12x(x^6 + 4x^3) =$

25) $5x(3x^2 + 6x + 4) =$

26) $7x(4 - 2x + 6x^5) =$

27) $3x(4x^4 - 4x^3 + 2) =$

28) $4x(2x^5 + 6x^2 - 3) =$

29) $5x(3x^4 + 4x^3 + 2x) =$

30) $2x(3x - 2x^3 + 4x^6) =$

Simplifying Polynomials

Simplify each expression.

1) $3(4x-20)=$

2) $5x(3x-4)=$

3) $6x(5x-7)=$

4) $3x(7x+5)=$

5) $5x(4x-3)=$

6) $6x(8x+2)=$

7) $(3x-2)(x-4)=$

8) $(x-5)(2x+6)=$

9) $(x-3)(x-7)=$

10) $(3x+4)(3x-4)=$

11) $(5x-4)(5x-2)=$

12) $6x^2+6x^2-8x^4=$

13) $3x-2x^2+5x^3+7=$

14) $7x+4x^2-10x^3=$

15) $12x^2+5x^5-6x^3=$

16) $-5x^2+4x^6+6x^8=$

17) $-12x^3+10x^5-4x^6+4x=$

18) $11-7x^2+4x^2-16x^3+11=$

19) $2x^2-9x+4x^3+15x-10x=$

20) $13-7x^5+6x^5-4x^2+5=$

21) $-5x^8+x^6-14x^3+5x^8=$

22) $(7x^4-4)+(7x^4-2x^4)=$

23) $3(3x^4-4x^3-6x^4)=$

24) $-5(x^9+8)-5(10-x^9)=$

25) $8x^3-9x^4-2x+19-8x^3=$

26) $11-8x^3+6x^3-7x^5+6=$

27) $(5x^3-4x)-(6x-2-6x^3)=$

28) $4x^2-5x^4-x(3x^3+2x)=$

29) $6x+6x^5-10-4(x^5-3)=$

30) $4-3x^4+(6x^5-2x^4+5x^5)=$

31) $-(x^5+4)-8(3+x^5)=$

32) $(4x^3-3x)-(3x-5x^3)=$

Adding and Subtracting Polynomials

Add or subtract expressions.

1) $(-2x^2 - 3) + (3x^2 + 4) =$

2) $(4x^3 + 6) - (7 - 2x^3) =$

3) $(4x^5 + 5x^2) - (2x^5 + 15) =$

4) $(6x^3 - 2x^2) + (5x^2 - 4x) =$

5) $(10x^4 + 28x) - (34x^4 + 6) =$

6) $(7x^2 - 3) + (7x^2 + 3) =$

7) $(9x^2 + 4) - (10 - 5x^2) =$

8) $(6x^2 + x^5) - (x^5 + 4) =$

9) $(4x^3 - x) + (3x - 7x^3) =$

10) $(11x + 10) - (8x + 10) =$

11) $(15x^3 - 3x) - (3x - 4x^3) =$

12) $(4x - x^5) - (6x^5 + 8x) =$

13) $(2x^2 - 7x^7) - (4x^7 - 6x) =$

14) $(3x^2 - 5) + (8x^2 + 4x^5) =$

15) $(9x^4 + 5x^5) - (x^5 - 9x^4) =$

16) $(-4x^3 - 2x) + (9x - 5x^3) =$

17) $(4x - 3x^2) - (148x^2 + x) =$

18) $(5x - 8x^4) - (3x^4 - 4x^2) =$

19) $(8x^4 - 4) + (2x^4 - 3x^2) =$

20) $(5x^6 + 7x^3) - (x^3 - 5x^6) =$

21) $(-2x^2 + 20x^5 + 5x^4) + (12x^4 + 8x^5 + 24x^2) =$

22) $(7x^4 - 9x^7 - 6x) - (-3x^4 - 9x^7 + 6x) =$

23) $(14\text{x} + 12x^4 - 18x^6) + (20x^4 + 18x^6 - 10x) =$

24) $(5x^8 - 6x^6 - 4x) - (5x^3 + 9x^6 - 7x) =$

25) $(11x^2 - 6x^4 - 3x) - (-4x^2 - 12x^4 + 9x) =$

26) $(-5x^9 + 14x^3 + 3x^7) + (10x^7 + 26x^3 + 3x^9) =$

Multiplying Monomials

Simplify each expression.

1) $6u^8 \times (-u^2) =$

2) $(-5p^8) \times (-2p^3) =$

3) $4xy^3z^5 \times 3z^4 =$

4) $3u^5t \times 8ut^4 =$

5) $(-5a^2) \times (-7a^3b^6) =$

6) $-3a^4b^3 \times 6a^2b =$

7) $13xy^5 \times x^4y^4 =$

8) $6p^4q^3 \times (-8pq^6) =$

9) $8s^4t^3 \times 4st^3 =$

10) $(-6x^4y^3) \times 6x^2y =$

11) $3xy^7z \times 12z^3 =$

12) $24xy \times x^2y =$

13) $13pq^4 \times (-3p^2q) =$

14) $13s^3t^4 \times st^4 =$

15) $11p^5 \times (-6p^3) =$

16) $(-8p^3q^5r) \times 3pq^4r^6 =$

17) $(-4a^4) \times (-7a^3b) =$

18) $6u^6v^2 \times (-5u^3v^4) =$

19) $9u^5 \times (-3u) =$

20) $-6xy^5 \times 4x^2y =$

21) $13y^5z^3 \times (-y^3z) =$

22) $8a^4bc^3 \times 2abc^3 =$

23) $(-7p^5q^6) \times (-5p^4q^2) =$

24) $4u^5v^3 \times (-4u^7v^3) =$

25) $17y^4z^5 \times (-y^6z) =$

26) $(-5pq^3r^2) \times 8p^2q^4r =$

27) $3ab^5c^6 \times 5a^4bc^2 =$

28) $6x^3yz^2 \times 3x^2y^7z^3 =$

Multiplying and Dividing Monomials

Simplify each expression.

1) $(5x^5)(2x^2) =$

2) $(4x^4)(6x^2) =$

3) $(3x^4)(7x^4) =$

4) $(5x^6)(4x^2) =$

5) $(12x^4)(3x^6) =$

6) $(4yx^8)(8y^4x^3) =$

7) $(14x^4y)(x^3y^5) =$

8) $(-5x^3y^4)(2x^3y^5) =$

9) $(-6x^4y^2)(-3x^3y^5) =$

10) $(5x^3y)(-5x^2y^3) =$

11) $(6x^4y^3)(4x^3y^4) =$

12) $(4x^3y^2)(5x^2y^4) =$

13) $(12x^3y^6)(4x^4y^{10}) =$

14) $(15x^3y^5)(3x^4y^6) =$

15) $(7x^2y^7)(8x^6y^7) =$

16) $(-3x^3y^8)(7x^9y^4) =$

17) $\frac{5x^6y^6}{xy^4} =$

18) $\frac{19x^7y^5}{19x^6y} =$

19) $\frac{56x^4y^4}{8xy} =$

20) $\frac{81x^5y^6}{9x^4y^5} =$

21) $\frac{36x^7y^6}{9x^2y^3} =$

22) $\frac{48x^9y^7}{4x^4y^6} =$

23) $\frac{88x^{18}y^{12}}{11x^8y^9} =$

24) $\frac{30x^7y^6}{6x^8y^3} =$

25) $\frac{150x^7y^6}{30x^4y^6} =$

26) $\frac{-42x^{18}y^{14}}{6x^4y^9} =$

27) $\frac{-36x^7y^8}{9x^5y^8} =$

Multiplying a Polynomial and a Monomial

Find each product.

1) $x(2x+4)=$

2) $6(4-2x)=$

3) $5x(4x+2)=$

4) $x(-4x+5)=$

5) $8x(2x-2)=$

6) $6(2x-4y)=$

7) $7x(5x-5)=$

8) $3x(12x+2y)=$

9) $4x(x+6y)=$

10) $11x(3x+4y)=$

11) $7x(3x+2)=$

12) $10x(4x-10y)=$

13) $9x(3x-2y)=$

14) $7x(x-4y+6)=$

15) $8x(2x^2+5y^2)=$

16) $12x(2x+3y)=$

17) $4(2x^4-4y^4)=$

18) $4x(-3x^2y+4y)=$

19) $-4(5x^3-2xy+4)=$

20) $4(x^2-5xy-6)=$

21) $8x(2x^3-5xy+2x)=$

22) $-6x(-2x^3-6x+2xy)=$

23) $3(2x^2+xy-9y^2)=$

24) $4x(5x^3-3x+7)=$

25) $6(3x^{22}-2x-5)=$

26) $x^2(-2x^3+4x+3)=$

27) $x^2(4x^3+10-2x)=$

28) $4x^4(3x^3-2x+5)=$

29) $2x^2(4x^4-5xy+7y^3)=$

30) $5x^2(5x^4-3x+9)=$

31) $7x^2(6x^2+3x-6)=$

32) $4x(x^3-4xy+2y^2)=$

Multiplying Binomials

Find each product.

1) $(x+3)(x+6) =$

2) $(x-4)(x+3) =$

3) $(x-3)(x-8) =$

4) $(x+8)(x+9) =$

5) $(x-2)(x-12) =$

6) $(x+5)(x+5) =$

7) $(x-6)(x+7) =$

8) $(x-8)(x-3) =$

9) $(x+7)(x+12) =$

10) $(x-4)(x+8) =$

11) $(x+8)(x+8) =$

12) $(x+2)(x+7) =$

13) $(x-6)(x+6) =$

14) $(x-5)(x+5) =$

15) $(x+11)(x+11) =$

16) $(x+6)(x+9) =$

17) $(x-2)(x+2) =$

18) $(x-4)(x+7) =$

19) $(3x+5)(x+6) =$

20) $(5x-6)(4x+8) =$

21) $(x-7)(3x+7) =$

22) $(x-9)(x-4) =$

23) $(x-12)(x+2) =$

24) $(2x-4)(5x+4) =$

25) $(3x-8)(x+8) =$

26) $(7x-2)(6x+3) =$

27) $(4x+5)(3x+5) =$

28) $(7x-4)(9x+4) =$

29) $(x+2)(2x-8) =$

30) $(5x-4)(5x+4) =$

31) $(3x+2)(3x-7) =$

32) $(x^2+8)(x^2-8) =$

Factoring Trinomials

Factor each trinomial.

1) $x^2 + 8x + 12 =$

2) $x^2 - 6x + 5 =$

3) $x^2 + 15x + 36 =$

4) $x^2 - 12x + 35 =$

5) $x^2 - 11x + 18 =$

6) $x^2 - 9x + 18 =$

7) $x^2 + 18x + 72 =$

8) $x^2 - x - 72 =$

9) $x^2 + 4x - 21 =$

10) $x^2 - 13x + 22 =$

11) $x^2 + 2x - 24 =$

12) $x^2 - 3x - 40 =$

13) $x^2 - 3x - 70 =$

14) $x^2 + 26x + 169 =$

15) $4x^2 - 7x - 15 =$

16) $x^2 - 14x + 33 =$

17) $10x^2 + 5x - 15 =$

18) $6x^2 - 4x - 42 =$

19) $x^2 + 12x + 36 =$

20) $5x^2 + 17x - 12 =$

Calculate each problem.

21) The area of a rectangle is $x^2 - x - 56$. If the width of rectangle is $x + 7$, what is its length? ______________

22) The area of a parallelogram is $4x^2 + 17x - 15$ and its height is $x + 5$. What is the base of the parallelogram? ______________

23) The area of a rectangle is $6x^2 - 22x + 12$. If the width of the rectangle is $3x - 2$, what is its length? ______________

Operations with Polynomials

Find each product.

1) $4(5x+3) =$ ____________
2) $8(2x+6) =$ ____________
3) $2(5x-2) =$ ____________
4) $-4(7x-3) =$ ____________
5) $3x^2(9x+1) =$ ____________
6) $4x^6(7x-9) =$ ____________
7) $3x^4(-7x+3) =$ ____________
8) $-8x^4\,(5x-8) =$ ____________
9) $7\,(x^2+5x-3) =$ ____________
10) $9(5x^2-7x+5) =$ ____________
11) $3(3x^2+3x+2) =$ ____________
12) $5x(3x^2+5x+8) =$ ____________
13) $(5x+7)(3x-3) =$ ____________
14) $(9x+3)(3x-5) =$ ____________
15) $(6x+3)(4x-2) =$ ____________
16) $(7x-2)(3x+5) =$ ____________

Calculate each problem.

17) The measures of two sides of a triangle are $(2x+5y)$ and $(6x-3y)$. If the perimeter of the triangle is $(13x+4y)$, what is the measure of the third side? ____________

18) The height of a triangle is $(8x+5)$ and its base is $(4x-3)$. What is the area of the triangle? ____________

19) One side of a square is $(6x+2)$. What is the area of the square? ____________

20) The length of a rectangle is $(5x-8y)$ and its width is $(15x+8y)$. What is the perimeter of the rectangle? ____________

21) The side of a cube measures $(x+2)$. What is the volume of the cube? ____________

22) If the perimeter of a rectangle is $(28x+6y)$ and its width is $(5x+2y)$, what is the length of the rectangle? ____________

Answers of Worksheets

Writing Polynomials in Standard Form

1) $4x$
2) -5
3) $6x^5 - 12x^3$
4) $14x^4$
5) $-9x^3 + 5x^2 + 4x$
6) $12x^5 - 3x^2$
7) $-2x^8 + 8x^3 + 5x$
8) $-9x^6 - 7x^3 + 4x$
9) $3x^2 - 6x + 22$
10) $9x^4 - 4x + 3$
11) $-8x^3 + 13x^2 + 28x$
12) $-2x^3 + 4x^2 + 16$
13) $9x^4 + 19x^2 - 9x$
14) $3x^4 - 2x^3 - 7x^2$
15) $-8x^4 + 3x^2 - 51$
16) $-8x^6 + 4x^4 + 7x^2 - 15$
17) $-4x^5 + 6x^4 - 3x^3 + 16$
18) $-2x^6 - 7x^2 - x$
19) $6x^7 + 8x^5 - 3x^2$
20) $3x^6 - 12x^5 + 10x^2$
21) $-11x^7 + 20x^5 - 6x^3$
22) $-5x^5 + 34x^4 + 6x^3$
23) $-6x^3 + 12x^2 + 15x$
24) $12x^7 + 48x^4$
25) $15x^3 + 30x^2 + 20x$
26) $42x^6 - 14x^2 + 28x$
27) $12x^5 - 12x^4 + 6x$
28) $8x^6 + 24x^3 - 12x$
29) $15x^5 + 20x^4 + 10x^2$
30) $8x^7 - 4x^4 + 6x^2$

Simplifying Polynomials

1) $12x - 60$
2) $15x^2 - 20x$
3) $30x^2 - 42x$
4) $21x^2 + 15x$
5) $20x^2 - 15x$
6) $48x^2 + 12x$
7) $3x^2 - 14x + 8$
8) $2x^2 - 4x - 30$
9) $x^2 - 10x + 21$
10) $9x^2 - 16$
11) $25x^2 - 30x + 8$
12) $-8x^4 + 12x^2$
13) $5x^3 - 2x^2 + 3x + 7$
14) $-10x^3 + 4x^2 + 7x$
15) $5x^5 - 6x^3 + 12x^2$
16) $6x^8 + 4x^6 - 5x^2$
17) $-4x^6 + 10x^5 - 12x^3 + 4x$
18) $-16x^3 - 3x^2 + 22$
19) $4x^3 + 2x^2 - 4x$
20) $-x^5 - 4x^2 + 18$
21) $x^6 - 14x^3$
22) $12x^4 - 4$
23) $-9x^4 - 12x^3$
24) -90

25) $-9x^4 - 2x + 19$
26) $-7x^5 - 2x^3 + 17$
27) $11x^3 - 10x + 2$
28) $-8x^4 + 2x^2$
29) $2x^5 + 6x + 2$
30) $11x^5 - 5x^4 + 4$
31) $-9x^5 - 28$
32) $9x^3 - 6x$

Adding and Subtracting Polynomials

1) $x^2 + 1$
2) $6x^3 - 1$
3) $2x^5 + 5x^2 - 15$
4) $6x^3 + 3x^2 - 4x$
5) $-24x^4 + 28x - 6$
6) $14x^2$
7) $14x^2 - 6$
8) $6x^2 - 4$
9) $-3x^3 + 2x$
10) $3x$
11) $19x^3 - 6x$
12) $-7x^5 - 4x$
13) $-11x^7 + 2x^2 + 6x$
14) $4x^5 + 11x^2 - 5$
15) $4x^5 + 18x^4$
16) $-9x^3 + 7x$
17) $-151x^2 + 3x$
18) $-11x^4 + 4x^2 + 5x$
19) $10x^4 - 3x^2 - 4$
20) $10x^6 + 6x^3$
21) $28x^5 + 17x^4 + 22x^2$
22) $10x^4 - 12x$
23) $32x^4 + 4x$
24) $5x^8 - 15x^6 - 5x^3 + 3x$
25) $6x^4 + 15x^2 - 12x$
26) $-2x^9 + 13x^7 + 40x^3$

Multiplying Monomials

1) $-6u^{10}$
2) $10p^{11}$
3) $12xy^3z^9$
4) $24u^6t^5$
5) $35a^5b^6$
6) $-18a^6b^4$
7) $13x^5y^9$
8) $-48p^5q^9$
9) $32s^5t^6$
10) $-36x^6y^4$
11) $36xy^7z^4$
12) $24px^3y^2$
13) $-39p^3q^5$
14) $13s^4t^8$
15) $-66p^8$
16) $-24p^4q^9r^7$
17) $28a^7b$
18) $-30u^9v^6$
19) $-27u^6$
20) $-24x^3y^6$
21) $-13y^8z^4$
22) $16a^5b^2c^6$
23) $35p^9q^8$
24) $-16u^{12}v^6$
25) $-17y^{10}z^6$
26) $-40p^3q^7r^3$
27) $15a^5b^6c^8$
28) $18x^5y^8z^5$

Multiplying and Dividing Monomials

1) $10x^7$
2) $24x^6$
3) $21x^8$
4) $20x^8$
5) $36x^{10}$
6) $32x^{11}y^5$
7) $14x^7y^6$
8) $-10x^6y^9$
9) $18x^7y^7$
10) $-25x^5y^4$
11) $24x^7y^7$
12) $20x^5y^6$

13) $48x^7y^{16}$
14) $45x^7y^{11}$
15) $56x^8y^{14}$
16) $-21x^{12}y^{12}$
17) $5x^5y^2$
18) xy^4
19) $7x^3y^3$
20) $9xy$
21) $4x^5y^3$
22) $12x^5y$
23) $8x^{10}y^3$
24) $5x^{-1}y^3$
25) $5x^3$
26) $-7x^{14}y^5$
27) $-4x^2$

Multiplying a Polynomial and a Monomial

1) $2x^2 + 4x$
2) $-12x + 24$
3) $20x^2 + 10x$
4) $-4x^2 + 5x$
5) $16x^2 - 16x$
6) $12x - 24y$
7) $35x^2 - 35x$
8) $36x^2 + 6xy$
9) $4x^2 + 24xy$
10) $33x^2 + 44xy$
11) $21x^2 + 14x$
12) $40x^2 - 100xy$
13) $27x^2 - 18xy$
14) $7x^2 - 28xy + 42x$
15) $16x^3 + 40xy^2$
16) $24x^2 + 36xy$
17) $8x^4 - 16y^4$
18) $-12x^3y + 16xy$
19) $-20x^3 + 8xy - 16$
20) $4x^2 - 20xy - 24$
21) $16x^4 - 40x^2y + 16x^2$
22) $12x^4 + 36x^2 - 12x^2y$
23) $6x^2 + 3xy - 27y^2$
24) $20x^4 - 12x^2 + 28x$
25) $18x^{22} - 12x - 30$
26) $-2x^5 + 4x^3 + 3x^2$
27) $4x^5 - 2x^3 + 10x^2$
28) $12x^7 - 8x^5 + 20x^4$
29) $8x^6 - 10x^3y + 14x^2y^3$
30) $25x^6 - 15x^3 + 45x^2$
31) $42x^4 + 21x^3 - 42x^2$
32) $4x^4 - 16x^2y + 8xy^2$

Multiplying Binomials

1) $x^2 + 9x + 18$
2) $x^2 - x - 12$
3) $x^2 - 11x + 24$
4) $x^2 + 17x + 72$
5) $x^2 - 14x + 24$
6) $x^2 + 10x + 25$
7) $x^2 + x - 42$
8) $x^2 - 11x + 24$
9) $x^2 + 19x + 84$
10) $x^2 + 4x - 32$
11) $x^2 + 16x + 64$
12) $x^2 + 9x + 14$
13) $x^2 - 36$
14) $x^2 - 25$

15) $x^2 + 22x + 121$
16) $x^2 + 15x + 54$
17) $x^2 - 4$
18) $x^2 + 3x - 28$
19) $3x^2 + 23x + 30$
20) $20x^2 + 16x - 48$
21) $3x^2 - 14x - 49$
22) $x^2 - 13x + 36$
23) $x^2 - 10\text{x} - 24$
24) $10x^2 - 12x - 16$
25) $3x^2 + 16x - 64$
26) $42x^2 + 9x - 6$
27) $12x^2 + 35x + 25$
28) $63x^2 - 8x - 16$
29) $2x^2 - 4x - 16$
30) $25x^2 - 16$
31) $9x^2 - 15x - 14$
32) $x^4 - 64$

Factoring Trinomials

1) $(x + 6)(x + 2)$
2) $(x - 5)(x - 1)$
3) $(x + 12)(x + 3)$
4) $(x - 5)(x - 7)$
5) $(x - 2)(x - 9)$
6) $(x - 6)(x - 3)$
7) $(x + 6)(x + 12)$
8) $(x + 8)(x - 9)$
9) $(x - 3)(x + 7)$
10) $(x - 11)(x - 2)$
11) $(x - 4)(x + 6)$
12) $(x - 8)(x + 5)$
13) $(x + 7)(x - 10)$
14) $(x + 13)(x + 13)$
15) $(4x + 5)(x - 3)$
16) $(x - 11)(x - 3)$
17) $(5x - 5)(2x + 3)$
18) $(2x - 6)(3x + 7)$
19) $(x + 6)(x + 6)$
20) $(5x - 3)(x + 4)$
21) $(x - 8)$
22) $(4x - 3)$
23) $(2x - 6)$

Operations with Polynomials

1) $20x + 12$
2) $16x + 48$
3) $10x - 4$
4) $-28x + 12$
5) $27x^3 + 3x^2$
6) $28x^7 - 36x^6$
7) $-21x^5 + 9x^4$
8) $-40x^5 + 64x^4$
9) $7x^2 + 35x - 21$
10) $45x^2 - 63x + 45$
11) $9x^2 + 9x + 6$
12) $15x^3 + 25x^2 + 40x$
13) $15x^2 + 6x - 21$
14) $27x^2 - 36x - 15$
15) $24x^2 - 6$
16) $21x^2 + 29x - 10$
17) $(5x + 2y)$
18) $16x^2 - 2x - \frac{15}{2}$
19) $36x^2 + 24x + 4$
20) $40x$
21) $x^3 + 6x^2 + 12x + 8$
22) $(9x + y)$

Chapter 9 :

Functions Operations and Quadratic

Topics that you will practice in this chapter:

- ✓ Evaluating Function
- ✓ Adding and Subtracting Functions
- ✓ Multiplying and Dividing Functions
- ✓ Composition of Functions
- ✓ Quadratic Equation
- ✓ Solving Quadratic Equations
- ✓ Quadratic Formula and the Discriminant
- ✓ Quadratic Inequalities
- ✓ Graphing Quadratic Functions
- ✓ Domain and Range of Radical Functions
- ✓ Solving Radical Equations

It's fine to work on any problem, so long as it generates interesting mathematics along the way – even if you don't solve it at the end of the day." – Andrew Wiles

Evaluating Function

Write each of following in function notation.

1) $h = -8x + 3$

2) $k = 2a - 14$

3) $d = 11t$

4) $y = \frac{5}{12}x - \frac{7}{12}$

5) $m = 24n - 210$

6) $c = p^2 - 5p + 10$

Evaluate each function.

7) $f(x) = 2x - 7$, find $f(-3)$

8) $g(x) = \frac{1}{9}x + 12$, find $f(18)$

9) $h(x) = -4x + 9$, find $f(3)$

10) $f(x) = -x + 19$, find $f(-3)$

11) $f(a) = 7a - 12$, find $f(3)$

12) $h(x) = 14 - 3x$, find $f(-4)$

13) $g(n) = 6n - 10$, find $f(2)$

14) $f(x) = -11x - 4$, find $f(-1)$

15) $k(n) = -20 - 3.5n$, find $f(2)$

16) $f(x) = -0.7x + 3.3$, find $f(-7)$

17) $g(n) = \frac{11n+8}{n}$, find $g(2)$

18) $g(n) = \sqrt{3n} + 12$, find $g(3)$

19) $h(x) = x^{-2} - 7$, find $h(\frac{1}{9})$

20) $h(n) = n^{-3} + 11$, find $h(\frac{1}{4})$

21) $h(n) = n^3 - 2$, find $h(\frac{1}{2})$

22) $h(n) = n^2 - 4$, find $h(-\frac{1}{3})$

23) $h(n) = 4n^2 - 13$, find $h(-5)$

24) $h(n) = -2n^3 - 6n$, find $h(2)$

25) $g(n) = \sqrt{16n^2} - \sqrt{n}$, find $g(4)$

26) $h(a) = \frac{-14a+9}{3a}$, find $h(-b)$

27) $k(a) = 12a - 14$, find $k(a-3)$

28) $h(x) = \frac{1}{9}x + 18$, find $h(-18x)$

29) $h(x) = 8x^2 + 16$, find $h(\frac{x}{2})$

30) $h(x) = x^4 - 20$, find $h(-2x)$

Adding and Subtracting Functions

Perform the indicated operation.

1) $f(x) = 2x + 3$

$g(x) = x + 7$

Find $(f - g)(2)$

2) $g(a) = -5a - 8$

$f(a) = -3a - 5$

Find $(g - f)(-2)$

3) $h(t) = 4t + 3$

$g(t) = 4t + 7$

Find $(h - g)(t)$

4) $g(a) = -6a - 10$

$f(a) = 3a^2 + 9$

Find $(g - f)(x)$

5) $g(x) = \frac{5}{6}x - 23$

$h(x) = \frac{5}{12}x + 25$

Find $g(12) - h(12)$

6) $h(x) = \sqrt{3x} - 2$

$g(x) = \sqrt{3x} + 5$

Find $(h + g)(12)$

7) $f(x) = x^{-1}$

$g(x) = x^2 + \frac{5}{x}$

Find $(f - g)(-3)$

8) $h(n) = n^2 + 2$

$g(n) = -4n + 6$

Find $(h - g)(2a)$

9) $g(x) = -2x^2 - 5 - 4x$

$f(x) = 7 + 2x$

Find $(g - f)(3x)$

10) $g(t) = 11t - 4$

$f(t) = -2t^2 + 5$

Find $(g + f)(-t)$

11) $f(x) = 8x + 9$

$g(x) = -5x^2 + 3x$

Find $(f - g)(-x^2)$

12) $f(x) = -3x^4 - 5x$

$g(x) = 2x^4 + 5x + 22$

Find $(f + g)(3x^2)$

Multiplying and Dividing Functions

Perform the indicated operation.

1) $g(x) = -2x - 1$

$f(x) = 4x + 3$

Find $(g.f)(2)$

2) $f(x) = 5x$

$h(x) = -2x + 3$

Find $(f.h)(-2)$

3) $g(a) = 5a - 2$

$h(a) = 2a - 3$

Find $(g.h)(-3)$

4) $f(x) = 2x - 7$

$h(x) = x - 5$

Find $(\frac{f}{h})(4)$

5) $f(x) = 8a^2$

$g(x) = 3 + 2a$

Find $(\frac{f}{g})(2)$

6) $g(a) = \sqrt{4a} + 2$

$f(a) = (-a)^4 + 1$

Find $(\frac{g}{f})(1)$

7) $g(t) = t^3 + 1$

$h(t) = 5t - 2$

Find $(g.h)(-2)$

8) $g(n) = n^2 + 2n - 4$

$h(n) = -5n + 3$

Find $(g.h)(1)$

9) $g(a) = (a - 3)^2$

$f(a) = a^2 + 4$

Find $(\frac{g}{f})(3)$

10) $g(x) = -3x^2 + \frac{4}{5}x + 9$

$f(x) = x^2 - 24$

Find $(\frac{g}{f})(5)$

11) $f(x) = 2x^3 - 5x^2 + 1$

$g(x) = 3x - 1$

Find $(f.g)(x)$

12) $f(x) = 5x - 2$

$g(x) = x^3 - 2x$

Find $(f.g)(x^2)$

Composition of Functions

Using $f(x) = 2x - 5$ and $g(x) = -2x$, find:

1) $f(g(2)) =$

4) $g(f(5)) =$

2) $f(g(-1)) =$

5) $f(g(3)) =$

3) $g(f(-4)) =$

6) $g(f(0)) =$

Using $f(x) = -\frac{1}{4}x + \frac{3}{4}$ and $g(x) = 2x^2$, find:

7) $g(f(-2)) =$

10) $f(f(1)) =$

8) $g(f(4)) =$

11) $g(f(-4)) =$

9) $g(g(1)) =$

12) $f(g(x)) =$

Using $f(x) = -2x + 2$ and $g(x) = x + 1$, find:

13) $g(f(1)) =$

16) $f(g(-3)) =$

14) $f(f(0)) =$

17) $g(f(2)) =$

15) $f(g(-1)) =$

18) $f(g(x)) =$

Using $f(x) = \sqrt{x+9}$ and $g(x) = x - 9$, find:

19) $f(g(9)) =$

22) $f(f(7)) =$

20) $g(f(-9)) =$

23) $g(f(-5)) =$

21) $f(g(4)) =$

24) $g(g(0)) =$

Quadratic Equation

Multiply.

1) $(x-4)(x+6)=$ __________

2) $(x+5)(x+7)=$ __________

3) $(x-6)(x+8)=$ __________

4) $(x+2)(x-9)=$ __________

5) $(x-7)(x-8)=$ __________

6) $(3x+2)(x-3)=$ __________

7) $(4x-3)(x+2)=$ __________

8) $(4x-5)(x+1)=$ __________

9) $(7x+1)(x-6)=$ __________

10) $(5x+1)(3x-3)=$ ________

Factor each expression.

11) $x^2-2x-8=$ __________

12) $x^2+8x+15=$ __________

13) $x^2-2x-24=$ __________

14) $x^2-10x+21=$ __________

15) $x^2+10x+21=$ __________

16) $4x^2+9x+5=$ __________

17) $5x^2+13x-6=$ __________

18) $5x^2+17x-12=$ __________

19) $2x^2+7x+5=$ __________

20) $9x^2-21x+6=$ __________

Calculate each equation.

21) $(x+6)(x-3)=0$

22) $(x+1)(x+8)=0$

23) $(3x+6)(x+5)=0$

24) $(2x-2)(4x+8)=0$

25) $x^2+x+10=22$

26) $x^2+11x+36=12$

27) $2x^2+9x+9=5$

28) $x^2+3x-24=4$

29) $5x^2+5x-40=20$

30) $8x^2+8x=48$

Solving Quadratic Equations

Solve each equation by factoring or using the quadratic formula.

1) $(x+9)(x-1)=0$

2) $(x+7)(x+6)=0$

3) $(x-8)(x+3)=0$

4) $(x-6)(x-4)=0$

5) $(x+2)(x+12)=0$

6) $(5x+4)(x+7)=0$

7) $(6x+1)(4x+5)=0$

8) $(2x+7)(x+8)=0$

9) $(x+6)(3x+15)=0$

10) $(12x+2)(x+8)=0$

11) $x^2=8x$

12) $x^2-16=0$

13) $3x^2+6=9x$

14) $-2x^2-8=10x$

15) $5x^2+40x=45$

16) $x^2+10x=24$

17) $x^2+6x=16$

18) $x^2+9x=-18$

19) $x^2+13x=-36$

20) $x^2+3x-15=5x$

21) $x^2+8x+7=-8$

22) $3x^2-11x=-9+x$

23) $10x^2+3=27x-15$

24) $7x^2-6x+8=8$

25) $2x^2-12=-3x+2$

26) $10x^2-26x-3=-15$

27) $3x^2+21=-16x+5$

28) $x^2+15x-10=-66$

29) $3x^2-8x-8=4+x$

30) $2x^2+6x-24=12$

31) $3x^2-33x+54=-18$

32) $-10x^2-15x-9=-9-27x^2$

Quadratic Formula and the Discriminant

Find the value of the discriminant of each quadratic equation.

1) $3x(x-8)=0$

2) $2x^2+6x-4=0$

3) $x^2+6x+7=0$

4) $x^2-x+3=0$

5) $x^2+4x-3=0$

6) $2x^2+6x-10=0$

7) $3x^2+7x+5=0$

8) $x^2-6x-4=0$

9) $2x^2+8x+3=0$

10) $x^2+7x-5=0$

11) $5x^2+2x-3=0$

12) $-3x^2-11x+4=0$

13) $-6x^2-12x+8=0$

14) $-x^2-9x-12=0$

15) $7x^2-6x-10=0$

16) $-4x^2-2x+8=0$

17) $5x^2+8x-2=0$

18) $6x^2-4x=0$

19) $3x^2-5x+2=0$

20) $4x^2+9x+3=0$

Find the discriminant of each quadratic equation then state the number of real and imaginary solutions.

21) $-4x^2-16=16x$

22) $20x^2=20x-5$

23) $-11x^2-19x=26$

24) $22x^2-4x+1=18x^2$

25) $-11x^2=-15x+8$

26) $3x^2+6x+9=6$

27) $13x^2-5x-12=-26$

28) $-8x^2-32x-25=7$

Graphing Quadratic Functions

Sketch the graph of each function. Identify the vertex and axis of symmetry.

1) $y = (x + 3)^2 + 2$

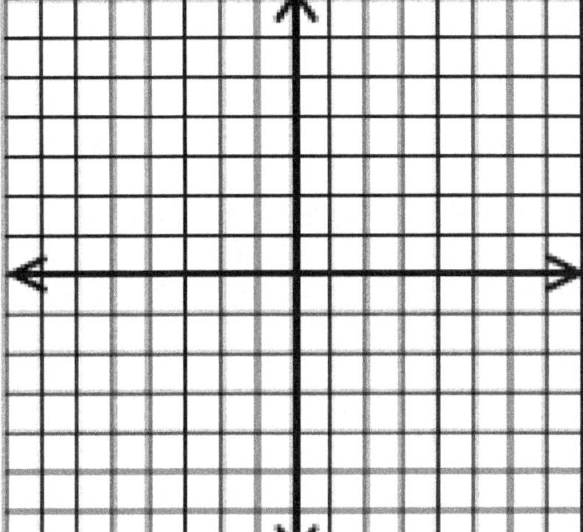

2) $y = (x - 3)^2 - 2$

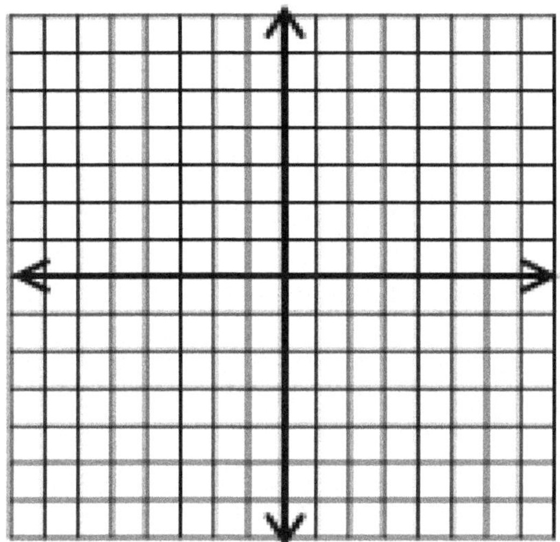

3) $y = 6 - (-x + 4)^2$

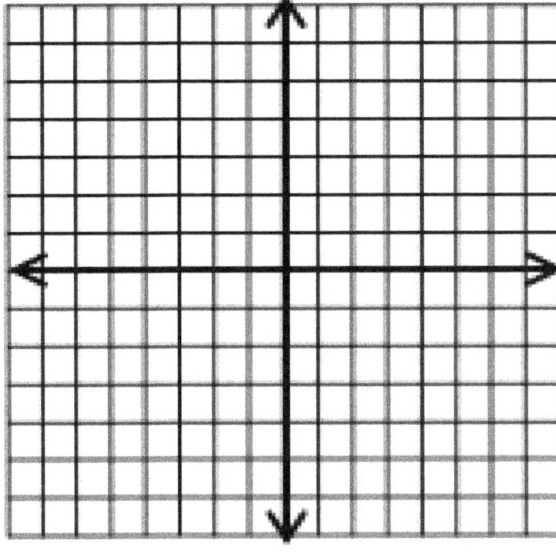

4) $y = -3x^2 - 6x + 9$

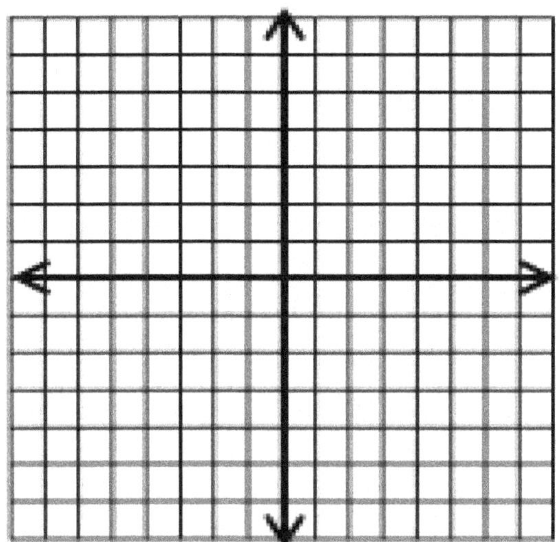

Quadratic Inequalities

Solve each quadratic inequality.

1) $x^2 - 25 < 0$

2) $-x^2 - 6x - 8 > 0$

3) $5x^2 + 15x + 30 < 0$

4) $x^2 + 8x + 16 > 0$

5) $2x^2 - 18x - 20 \geq 0$

6) $x^2 > -10x - 25$

7) $3x^2 + 2x + 16 \leq 0$

8) $x^2 - 5x - 14 \leq 0$

9) $x^2 - 6x - 7 \geq 0$

10) $2x^2 + 16x - 18 < 0$

11) $x^2 + 6x - 72 > 0$

12) $3x^2 - 3x - 36 > 0$

13) $x^2 - 15x + 64 \leq 0$

14) $2x^2 - 24x + 72 \leq 0$

15) $x^2 - 16x + 63 \geq 0$

16) $x^2 - 16x + 55 \geq 0$

17) $x^2 - 81 \leq 0$

18) $x^2 - 17x + 42 \geq 0$

19) $9x^2 + 14x + 36 \leq 0$

20) $4x^2 - 2x - 24 > 2x^2$

21) $5x^2 - 20x + 20 < 0$

22) $7x^2 - 6x \geq 6x^2 - 5$

23) $5x^2 - 15 > 4x^2 + 2x$

24) $3x^2 - 4x \geq 3x^2 - 9x + 15$

25) $8x^2 + 9x - 54 > 5x^2$

26) $10x^2 + 50x - 60 < 0$

27) $-x^2 + 15x - 57 \geq 0$

28) $-5x^2 + 25x + 30 \leq 0$

29) $5x^2 + 40x + 75 < 0$

30) $9x^2 + 20x + 180 \leq 0$

31) $3x^2 + 2x - 36 \geq -x$

32) $3x^2 + 9x + 9 \leq 6x^2 + 3x$

Domain and Range of Radical Functions

Identify the domain and range of each function.

1) $y = \sqrt{x+8} - 7$

2) $y = \sqrt[3]{3x-5} - 4$

3) $y = \sqrt{3x-9} + 3$

4) $y = \sqrt[3]{(4x+6)} - 2$

5) $y = 3\sqrt{4x+20} + 6$

6) $y = \sqrt[3]{(5x-2)} - 11$

7) $y = 4\sqrt{9x^2+8} + 3$

8) $y = \sqrt[3]{(7x^2-2)} - 6$

9) $y = 2\sqrt{2x^3+16} - 3$

10) $y = \sqrt[3]{(11x+4)} - 2x$

11) $y = 3\sqrt{-2(4x+8)} + 5$

12) $y = \sqrt[5]{(3x^2-12)} - 6$

13) $y = 3\sqrt{x-5} - 2$

14) $y = \sqrt[3]{6x+9} - 4$

Sketch the graph of each function.

15) $y = -3\sqrt{x} + 5$

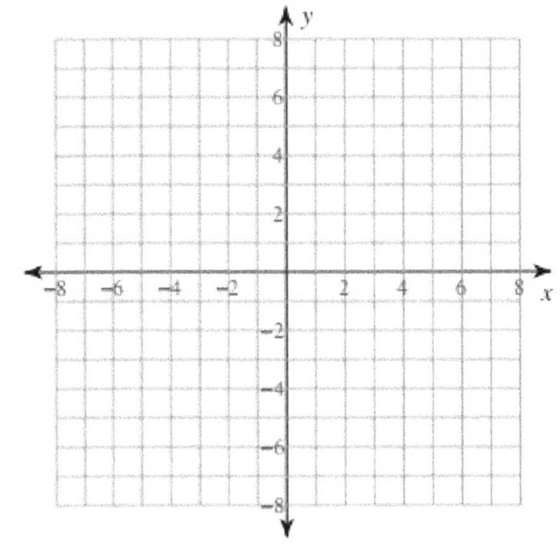

16) $y = 3\sqrt{x} - 6$

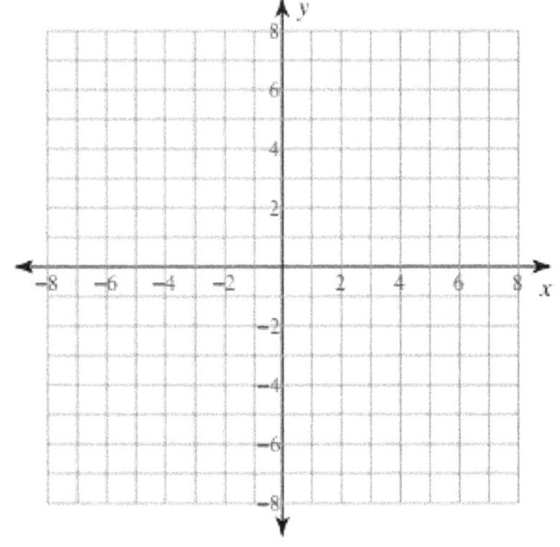

Solving Radical Equations

Solve each equation. Remember to check for extraneous solutions.

1) $\sqrt{a} = 9$

2) $\sqrt{v} = 6$

3) $\sqrt{r} = 4$

4) $8 = 16\sqrt{x}$

5) $\sqrt{x+3} = 18$

6) $6 = \sqrt{x-7}$

7) $4 = \sqrt{r-3}$

8) $\sqrt{x-5} = 7$

9) $12 = \sqrt{x-4}$

10) $\sqrt{m+5} = 8$

11) $7\sqrt{5a} = 35$

12) $6\sqrt{2x} = 48$

13) $2 = \sqrt{6x-32}$

14) $\sqrt{304-4x} = 4$

15) $\sqrt{r+2} - 8 = 6$

16) $-21 = -7\sqrt{r+9}$

17) $60 = 6\sqrt{5v}$

18) $x = \sqrt{40-3x}$

19) $\sqrt{90-27a} = 3a$

20) $\sqrt{-8n+88} = 4$

21) $\sqrt{15r-5} = 4r-3$

22) $\sqrt{-64+32x} = 4x$

23) $\sqrt{4x+15} = \sqrt{2x+11}$

24) $\sqrt{12v} = \sqrt{15v-21}$

25) $\sqrt{9-x} = \sqrt{x-3}$

26) $\sqrt{6m+34} = \sqrt{8m+34}$

27) $\sqrt{7r+32} = \sqrt{-8-3r}$

28) $\sqrt{4k+10} = \sqrt{2-4k}$

29) $-20\sqrt{x-13} = -40$

30) $\sqrt{90-2x} = \sqrt{\frac{x}{4}}$

Answers of Worksheets

Evaluating Function

1) $h(x) = -8x + 3$
2) $k(a) = 2a - 14$
3) $d(t) = 11t$
4) $f(x) = \frac{5}{12}x - \frac{7}{12}$
5) $m(n) = 24n - 210$
6) $c(p) = p^2 - 5p + 10$
7) -13
8) 14
9) -3
10) 22
11) 9
12) 26
13) 2
14) 7
15) -27
16) 8.2
17) 15
18) 15
19) 74
20) 75
21) $-1\frac{7}{8}$
22) $-3\frac{8}{9}$
23) 87
24) -28
25) 14
26) $-\frac{14b+9}{3b}$
27) $12a - 50$
28) $-2x + 18$
29) $2x^2 + 16$
30) $16x^4 - 20$

Adding and Subtracting Functions

1) -2
2) 1
3) -4
4) $-3x^2 - 6x - 19$
5) -43
6) 15
7) $-7\frac{2}{3}$
8) $4a^2 + 8a - 4$
9) $-18x^2 - 18x - 12$
10) $-2t^2 - 11t + 1$
11) $5x^4 - 5x^2 + 9$
12) $-81x^8 + 22$

Multiplying and Dividing Functions

1) -55
2) -70
3) 153
4) -1
5) $4\frac{4}{7}$
6) 2
7) 84
8) 2
9) 0
10) -62
11) $6x^4 - 17x^3 + 5x^2 + 3x - 1$
12) $5x^8 - 2x^6 - 10x^4 + 4x^2$

Composition of Functions

1) -13
2) -1
3) 26
4) -10
5) -17
6) 10
7) $\frac{25}{8}$
8) $\frac{1}{8}$
9) 8
10) $\frac{5}{8}$
11) $\frac{49}{8}$
12) $-\frac{1}{2}(x^2 - \frac{3}{2})$
13) 1
14) -2
15) 2
16) 6
17) -1
18) $-2x$

19) 3
20) -9
21) 2
22) $\sqrt{13}$
23) -7
24) -18

Quadratic Equations

1) $x^2 + 2x - 24$
2) $x^2 + 12x + 35$
3) $x^2 + 2x - 48$
4) $x^2 - 7x - 18$
5) $x^2 - 15x + 56$
6) $3x^2 - 7x - 6$
7) $4x^2 + 5x - 6$
8) $4x^2 - x - 5$
9) $7x^2 - 41x - 6$
10) $15x^2 - 12x - 3$
11) $(x - 4)(x + 2)$
12) $(x + 5)(x + 3)$
13) $(x - 6)(x + 4)$
14) $(x - 3)(x - 7)$
15) $(x + 3)(x + 7)$
16) $(4x + 5)(x + 1)$
17) $(5x - 2)(x + 3)$
18) $(5x - 3)(x + 4)$
19) $(2x + 5)(x + 1)$
20) $3(x - 2)(3x - 1)$
21) $x = -6, x = 3$
22) $x = -1, x = -8$
23) $x = -2, x = -5$
24) $x = 1, x = -2$
25) $x = 3, x = -4$
26) $x = -3, x = -8$
27) $x = -4, x = -\frac{1}{2}$
28) $x = 4, x = -7$
29) $x = 3, x = -4$
30) $x = -3, x = 2$

Solving quadratic equations

1) $\{-9, 1\}$
2) $\{-6, -7\}$
3) $\{8, -3\}$
4) $\{6, 4\}$
5) $\{-2, -12\}$
6) $\{-\frac{4}{5}, -7\}$
7) $\{-\frac{5}{4}, -\frac{1}{6}\}$
8) $\{-\frac{7}{2}, -8\}$
9) $\{-6, -5\}$
10) $\{-\frac{1}{6}, -8\}$
11) $\{8, 0\}$
12) $\{4, -4\}$
13) $\{2, 1\}$
14) $\{-4, -1\}$
15) $\{1, -9\}$
16) $\{2, -12\}$
17) $\{2, -8\}$
18) $\{-3, -6\}$
19) $\{-4, -9\}$
20) $\{5, -3\}$
21) $\{-5, -3\}$
22) $\{1, 3\}$
23) $\{\frac{6}{5}, \frac{3}{2}\}$
24) $\{\frac{6}{7}, 0\}$
25) $\{-\frac{7}{2}, 2\}$
26) $\{\frac{3}{5}, 2\}$
27) $\{-\frac{4}{3}, -4\}$
28) $\{-8, -7\}$
29) $\{4, -1\}$
30) $\{3, -6\}$
31) $\{3, 8\}$
32) $\{\frac{15}{17}, 0\}$

Quadratic formula and the discriminant

1) 576
2) 68
3) 8
4) -11
5) 28
6) 116
7) -11
8) 52
9) 40
10) 69
11) 64
12) 169
13) 336
14) 33
15) 316
16) 132
17) 104
18) 16
19) 1
20) 33
21) 0, *one real solution*
22) 0, *one real solution*
23) -783, *no solution*

24) $0, one\ real\ solution$

25) $-127, no\ solution$

26) $0, one\ real\ solution$

27) $-703, no\ solution$

28) $0, one\ real\ solution$

Graphing quadratic functions

1) $(-3, 2), x = -3$

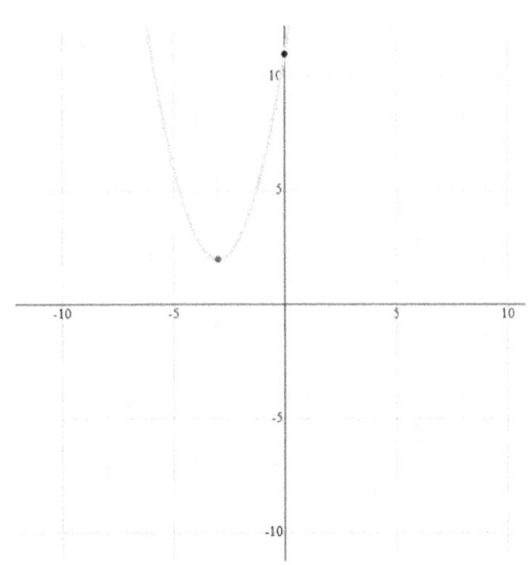

2) $(3, -2), x = 3$

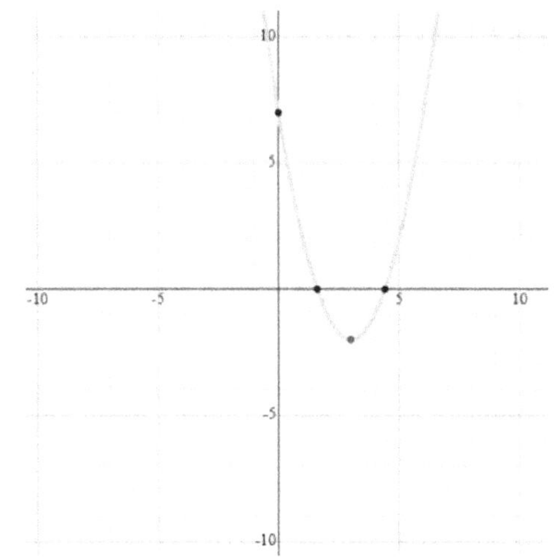

3) $(4, 6), x = 4$

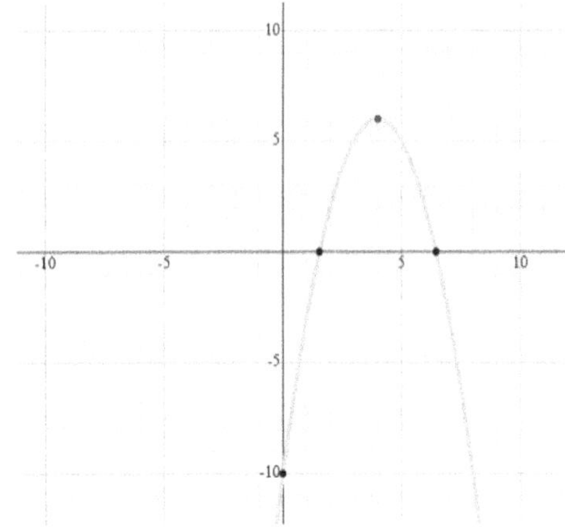

4) $(-1, 12), x = -1$

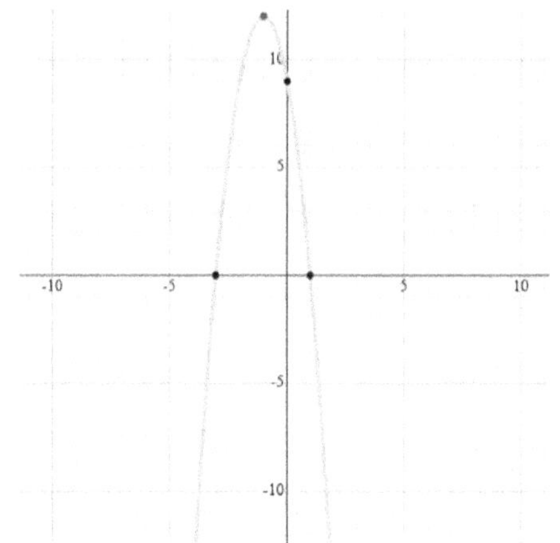

Quadratic inequalities

1) $-5 < x < 5$
2) $-4 < x < -2$
3) no solution
4) $x < -4\ or\ x > -4$
5) $x \leq -1\ or\ x \geq 10$
6) $x < -5\ or\ x > -5$
7) no solution
8) $-2 \leq x \leq 7$
9) $x \leq -1\ or\ x \geq 7$
10) $-9 < x < 1$
11) $x < -12\ or\ x > 6$
12) $-3 < x < 4$
13) no solution
14) $x = 6$
15) $x \leq 7 or\ \ x \geq 9$
16) $x \leq 5 or\ \ x \geq 11$
17) $-9 \leq x \leq 9$
18) $x \leq 3\ or\ x \geq 14$

19) no solution

20) $x < -3$ or $x > 4$

21) no solution

22) $x \leq 1$ or $x \geq 5$

23) $x < -3$ or $x > 5$

24) $x \geq 3$

25) $x < -6$ or $x > 3$

26) $-6 < x < 1$

27) no solution

28) $x \leq -1$ or $x \geq 6$

29) $-5 < x < -3$

30) no solution

31) $x \leq -4$ or $x \geq 3$

32) $x \leq -1$ or $x \geq 3$

Domain and range of radical functions

1) domain: $x \geq -8$

range: $y \geq -7$

2) domain: {all real numbers}

range: {all real numbers}

3) domain: $x \geq 3$

range: $y \geq 3$

4) domain: {all real numbers}

range: {all real numbers}

5) domain: $x \geq -5$

range: $y \geq 6$

6) domain: {all real numbers}

range: {all real numbers}

7) domain: {all real numbers}

range: $y \geq 8\sqrt{2} + 3$

8) domain: {all real numbers}

range: {all real numbers}

9) domain: $x \geq -2$

range: $y \geq -3$

10) domain: {all real numbers}

range: {all real numbers}

11) domain: $x \leq -2$

range: $y \geq 5$

12) domain: {all real numbers}

range: {all real numbers}

13) domain: $x \geq 5$

range: $y \geq -2$

14) domain: {all real numbers}

range: {all real numbers}

15)

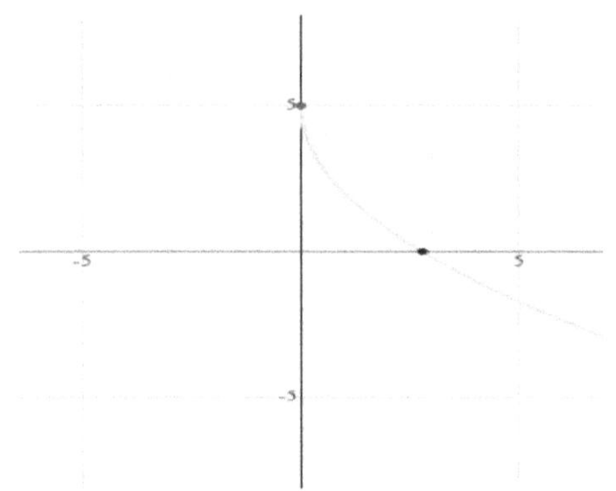

16)

Solving radical equations

1) $\{81\}$
2) $\{36\}$
3) $\{16\}$
4) $\{\frac{1}{4}\}$
5) $\{321\}$
6) $\{43\}$
7) $\{19\}$
8) $\{54\}$
9) $\{148\}$
10) $\{59\}$
11) $\{5\}$
12) $\{32\}$
13) $\{6\}$
14) $\{72\}$
15) $\{194\}$
16) $\{0\}$
17) $\{20\}$
18) $\{5\}$
19) $\{2\}$
20) $\{9\}$
21) $\{2\}$
22) no solution
23) $\{-2\}$
24) $\{7\}$
25) $\{6\}$
26) $\{0\}$
27) $\{-4\}$
28) $\{-1\}$
29) $\{17\}$
30) $\{40\}$

Chapter 10 :
Geometry and Solid Figures

Topics that you will practice in this chapter:

- ✓ Angles
- ✓ Pythagorean Relationship
- ✓ Triangles
- ✓ Polygons
- ✓ Trapezoids
- ✓ Circles
- ✓ Cubes
- ✓ Rectangular Prism
- ✓ Cylinder
- ✓ Pyramids and Cone

Mathematics is, as it were, a sensuous logic, and relates to philosophy as do the arts, music, and plastic art to poetry. — K. Shegel

Angles

What is the value of x in the following figures?

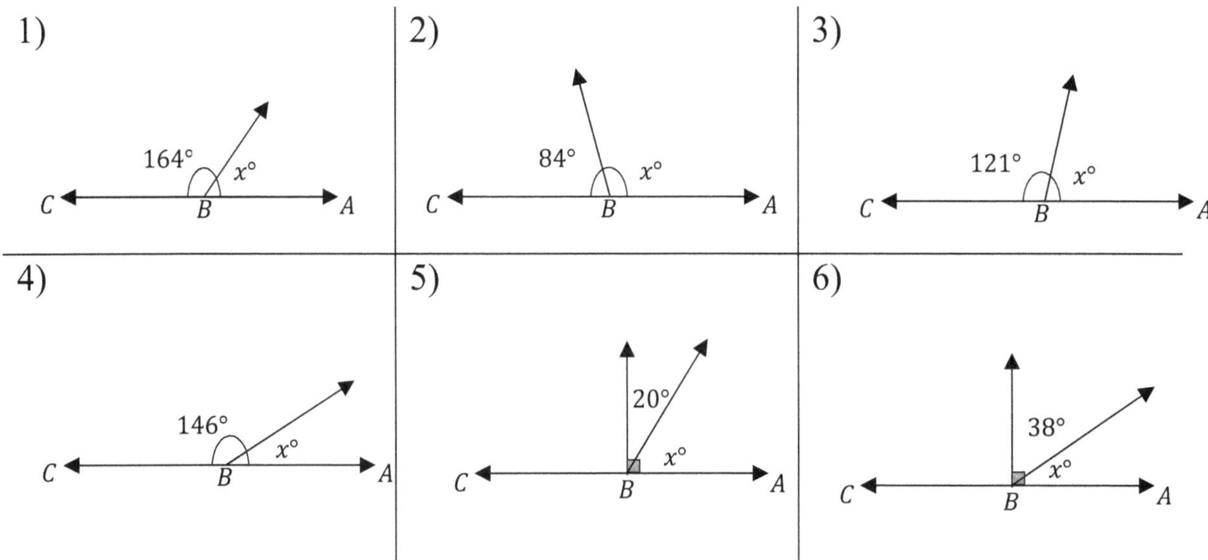

Calculate.

7) Two supplement angles have equal measures. What is the measure of each angle? ____________________

8) The measure of an angle is seven fifth the measure of its supplement. What is the measure of the angle? ____________________

9) Two angles are complementary and the measure of one angle is 24 less than the other. What is the measure of the smaller angle? ____________________

10) Two angles are complementary. The measure of one angle is one fifth the measure of the other. What is the measure of the bigger angle? ____________________

11) Two supplementary angles are given. The measure of one angle is 40° less than the measure of the other. What does the smaller angle measure? ____________________

Pythagorean Relationship

Do the following lengths form a right triangle?

1)

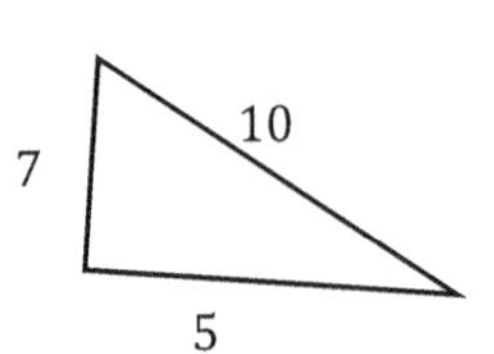

2)

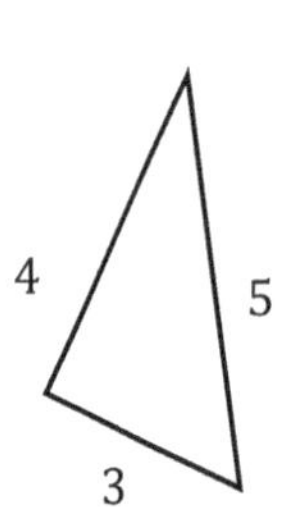

3)

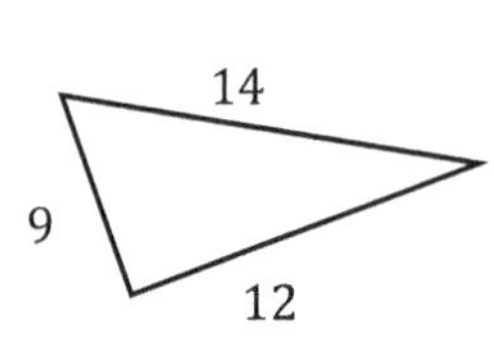

4)

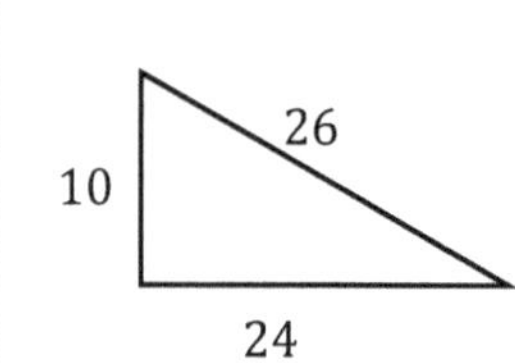

5)

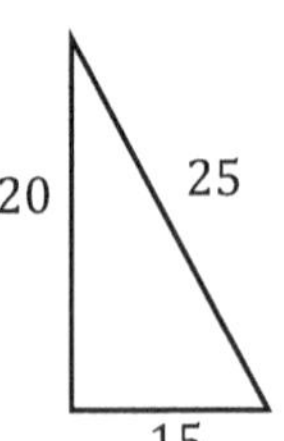

6)

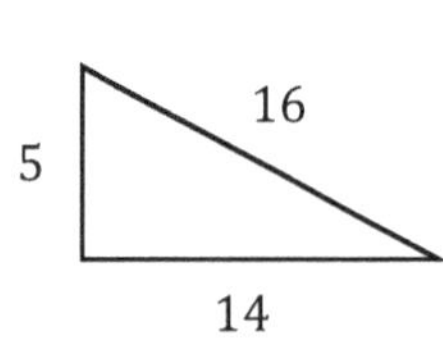

7)

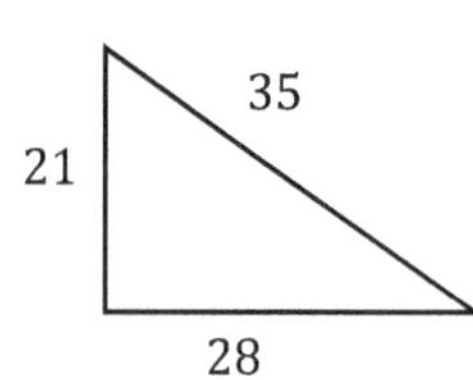

8)

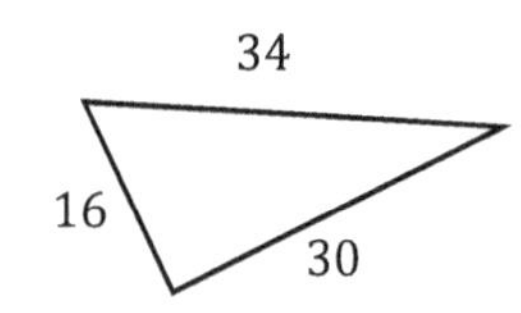

Find the missing side?

9)

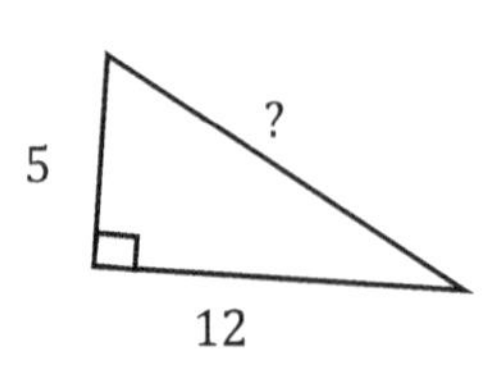

10)

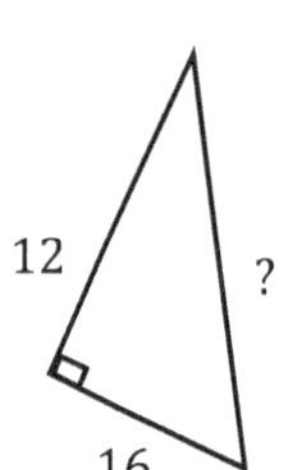

11)

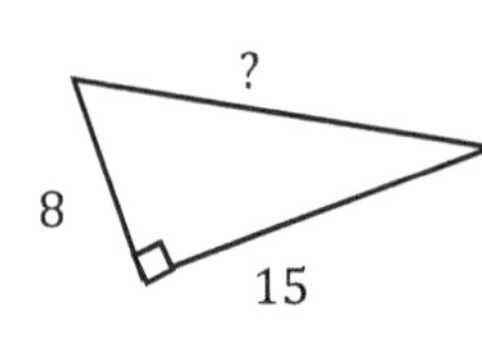

12)

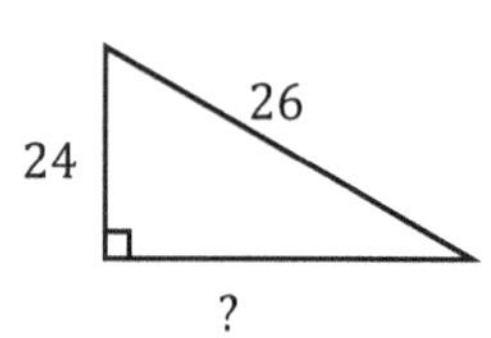

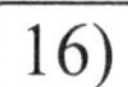

13)

14)

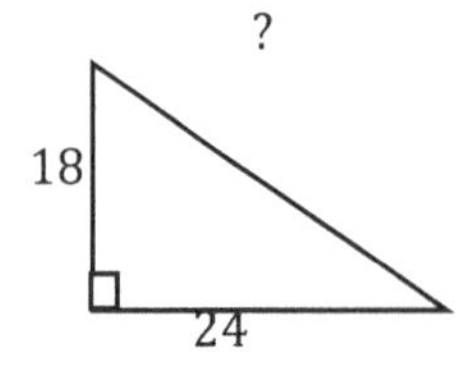

15)

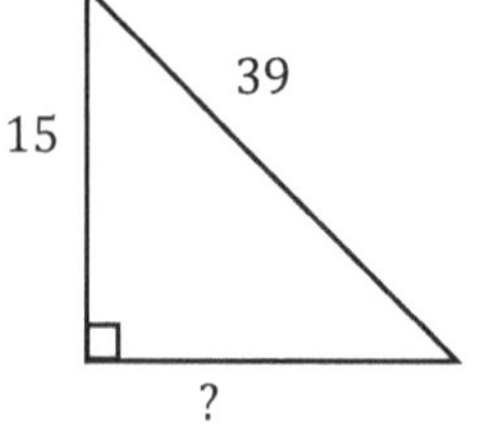

16)

15
9
?

Triangles

Find the measure of the unknown angle in each triangle.

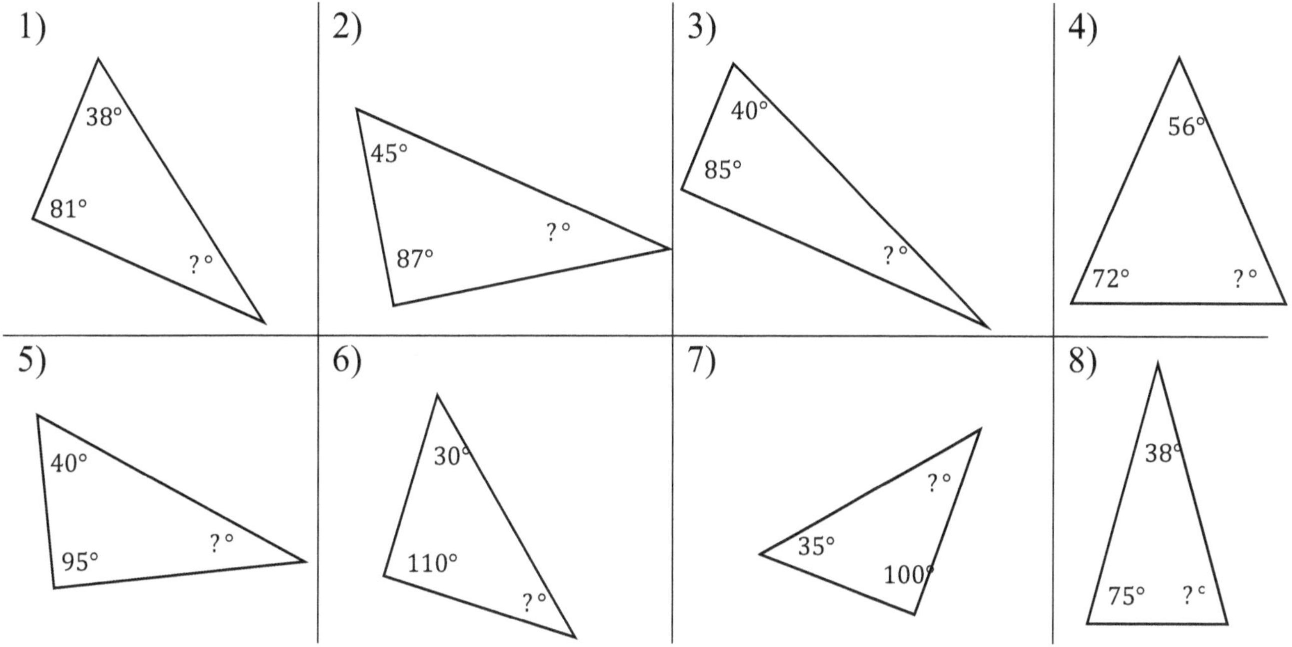

Find area of each triangle.

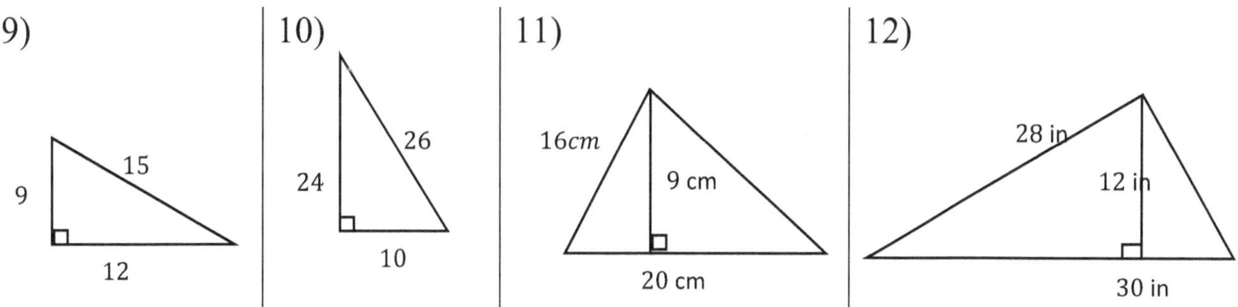

Polygons

Find the perimeter of each shape.

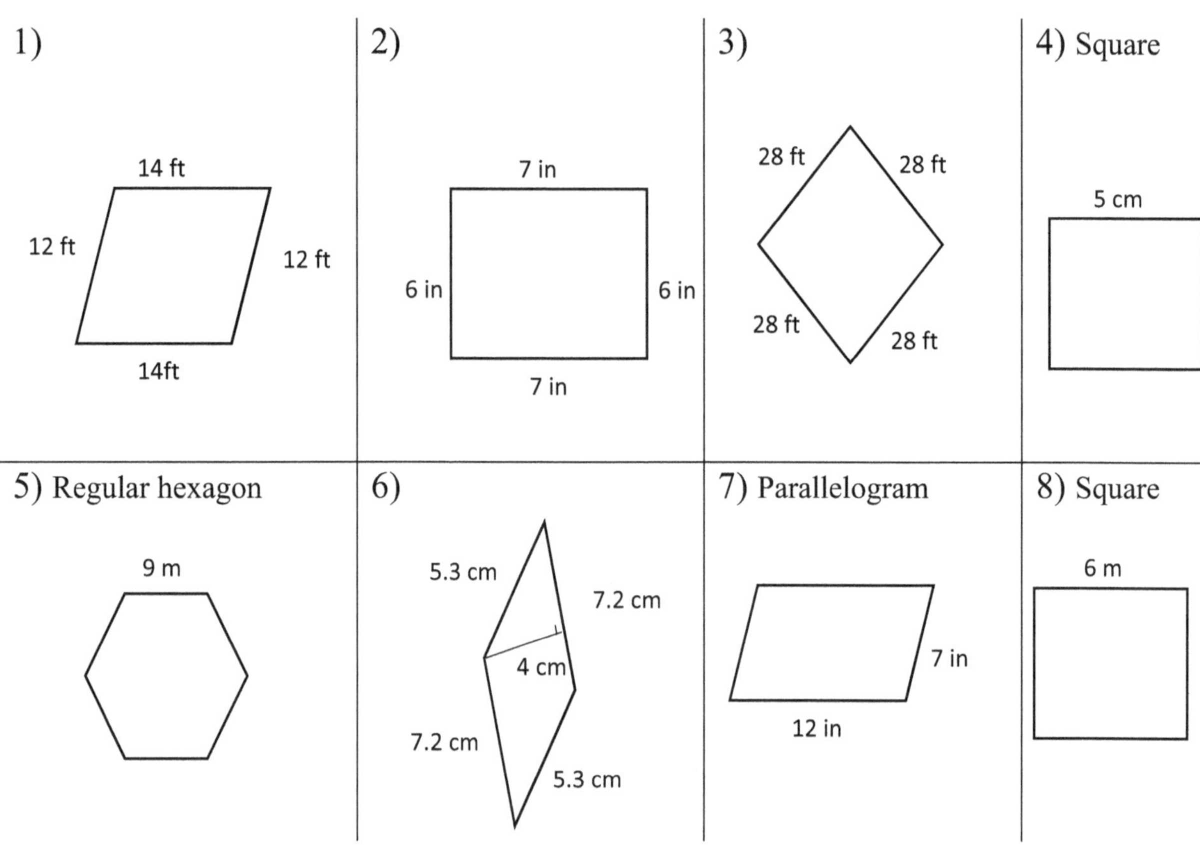

Find the area of each shape.

9) Parallelogram

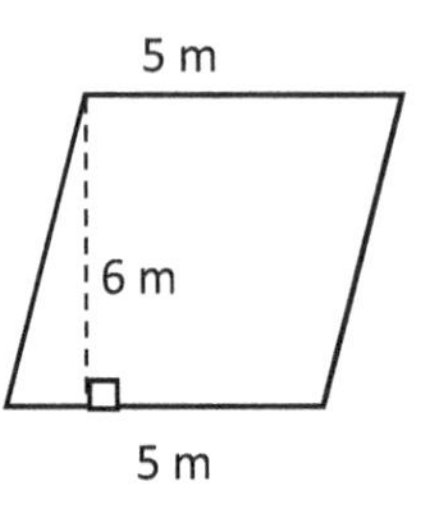

10) Rectangle

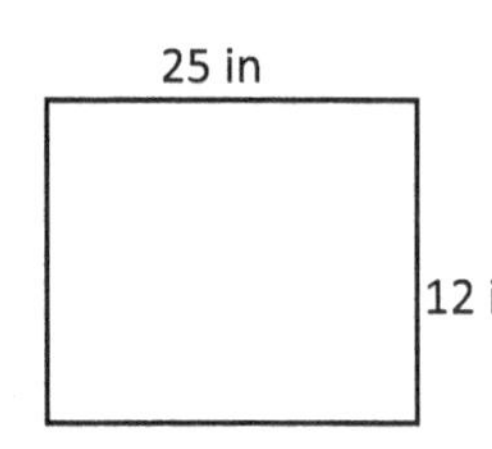

11) Rectangle

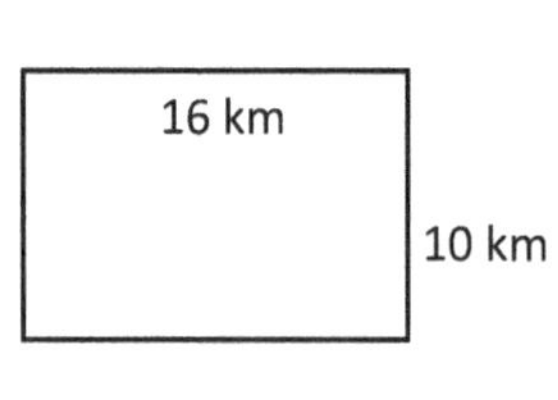

12) Square

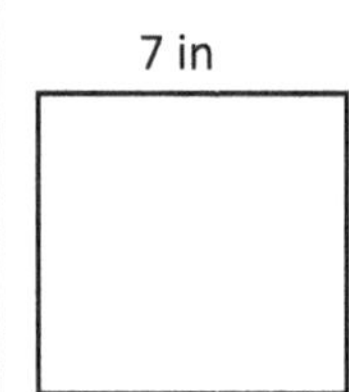

Trapezoids

Find the area of each trapezoid.

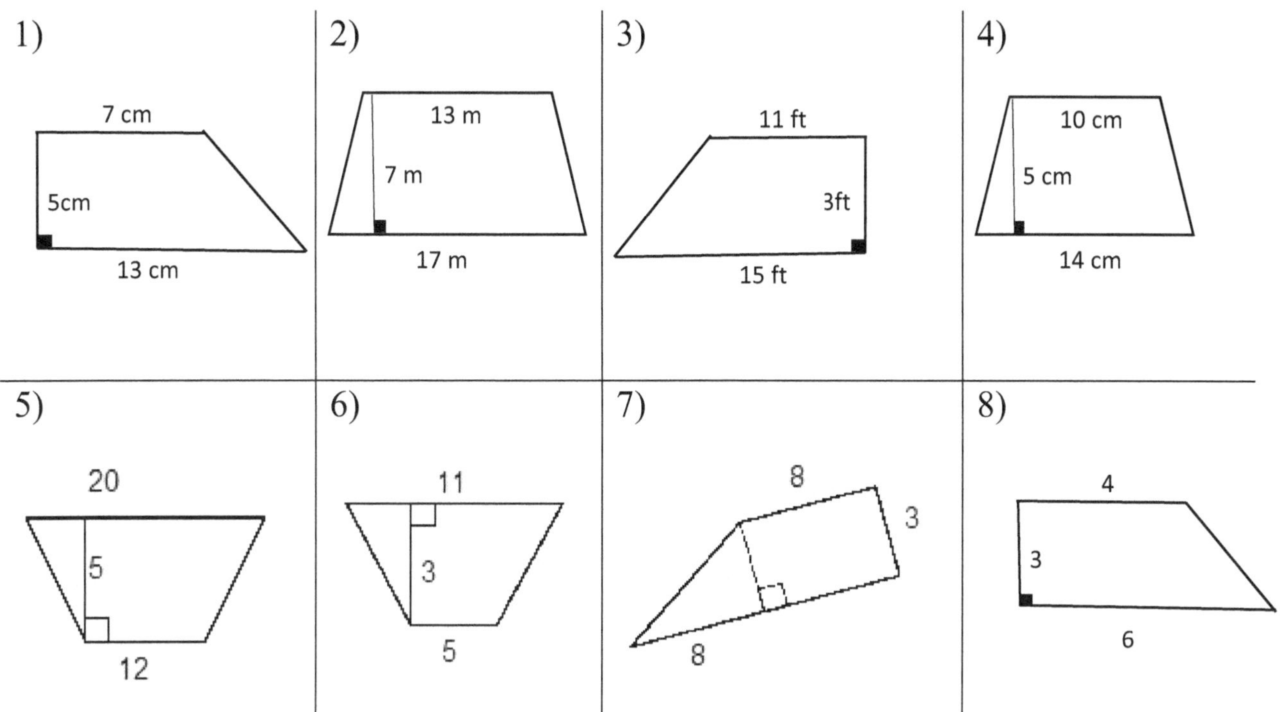

Calculate.

1) A trapezoid has an area of 45 cm^2 and its height is 5 cm and one base is 5 cm. What is the other base length? ____________________

2) If a trapezoid has an area of 99 ft^2 and the lengths of the bases are 8 ft and 10 ft, find the height? ____________________

3) If a trapezoid has an area of 126 m^2 and its height is 14 m and one base is 6 m, find the other base length? ____________________

4) The area of a trapezoid is 440 ft^2 and its height is 22 ft. If one base of the trapezoid is 15 ft, what is the other base length?

____

Circles

Find the area of each circle. ($\pi = 3.14$)

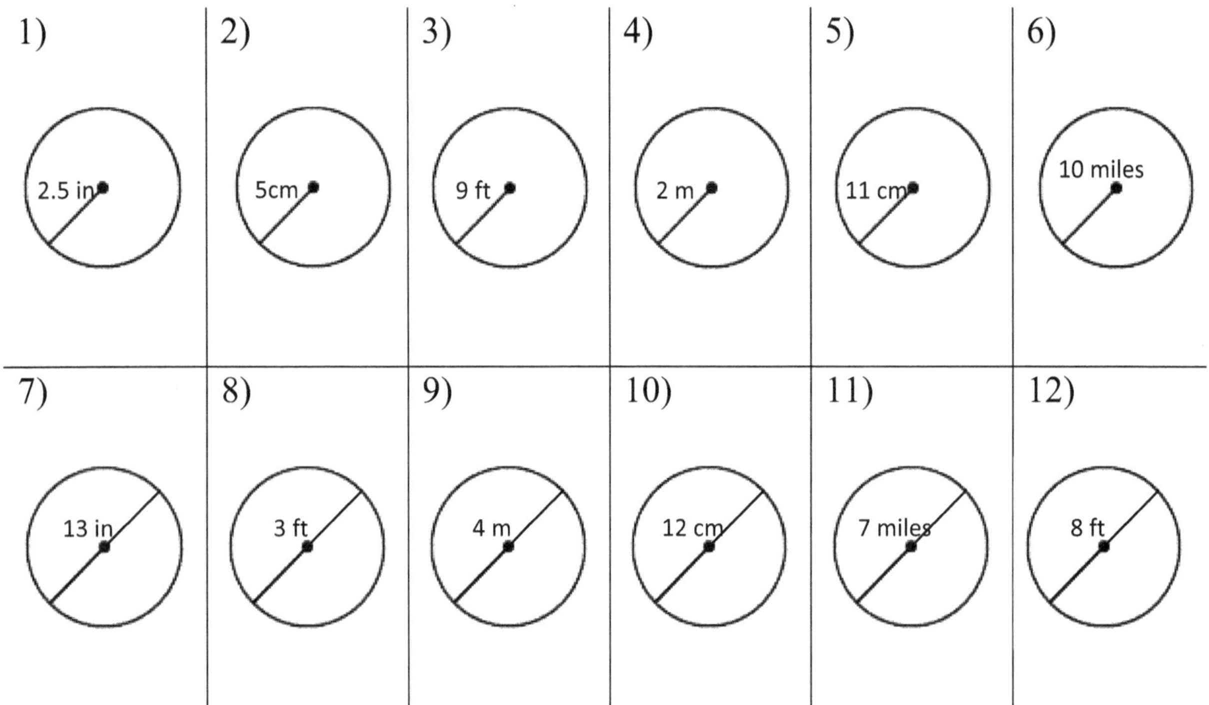

Complete the table below. ($\pi = 3.14$)

Circle No.	Radius	Diameter	Circumference	Area
1	1 in	2 in	6.28 in	3.14 in^2
2		10 m		
3				28.26 ft^2
4			47.1 mi	
5		11 km		
6	7 cm			
7		12 ft		
8				314 m^2
9			56.52 in	
10	4.5 ft			

Cubes

Find the volume of each cube.

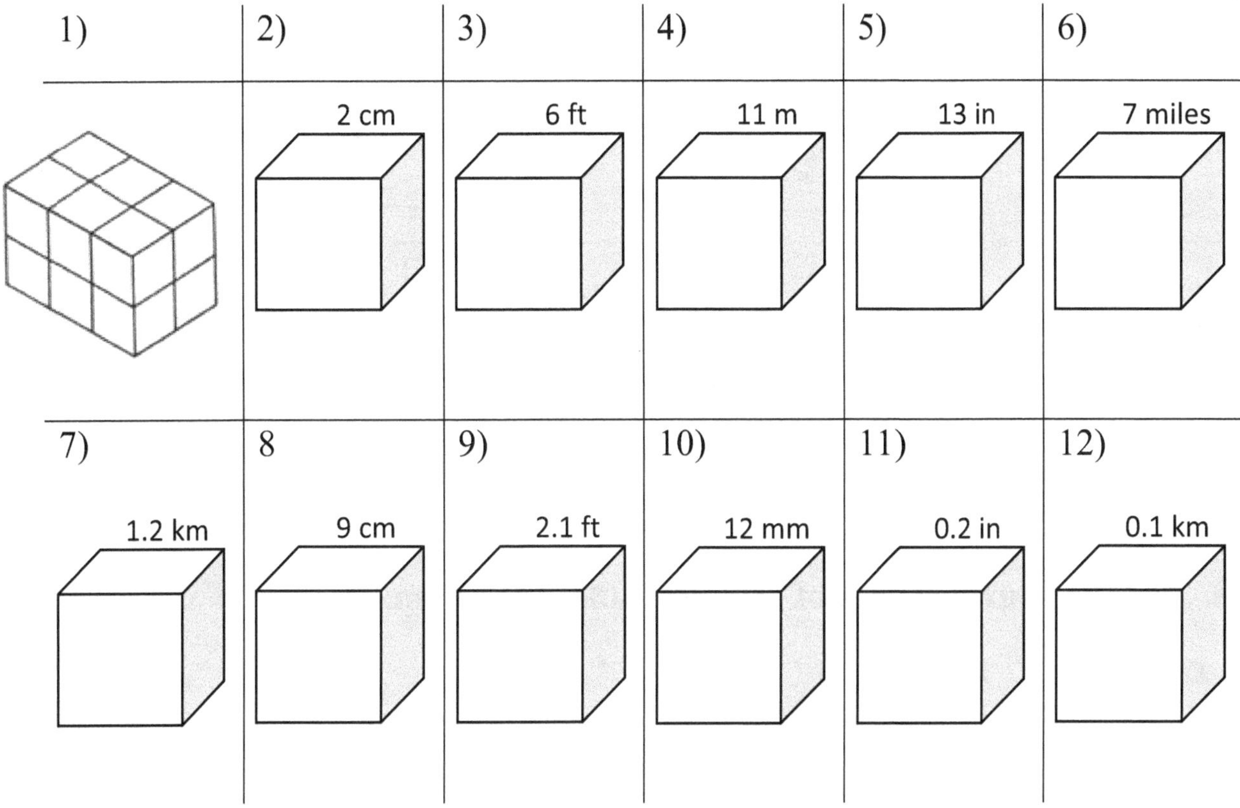

Find the surface area of each cube.

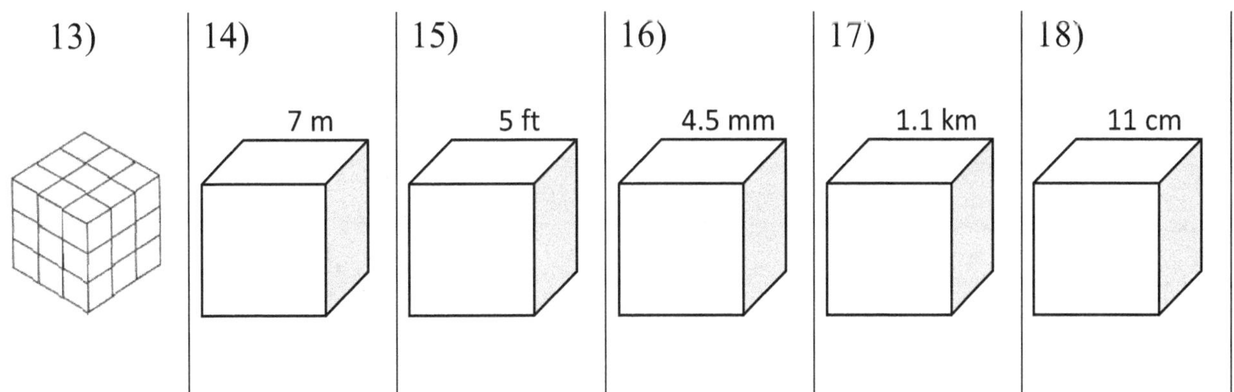

Rectangular Prism

Find the volume of each Rectangular Prism.

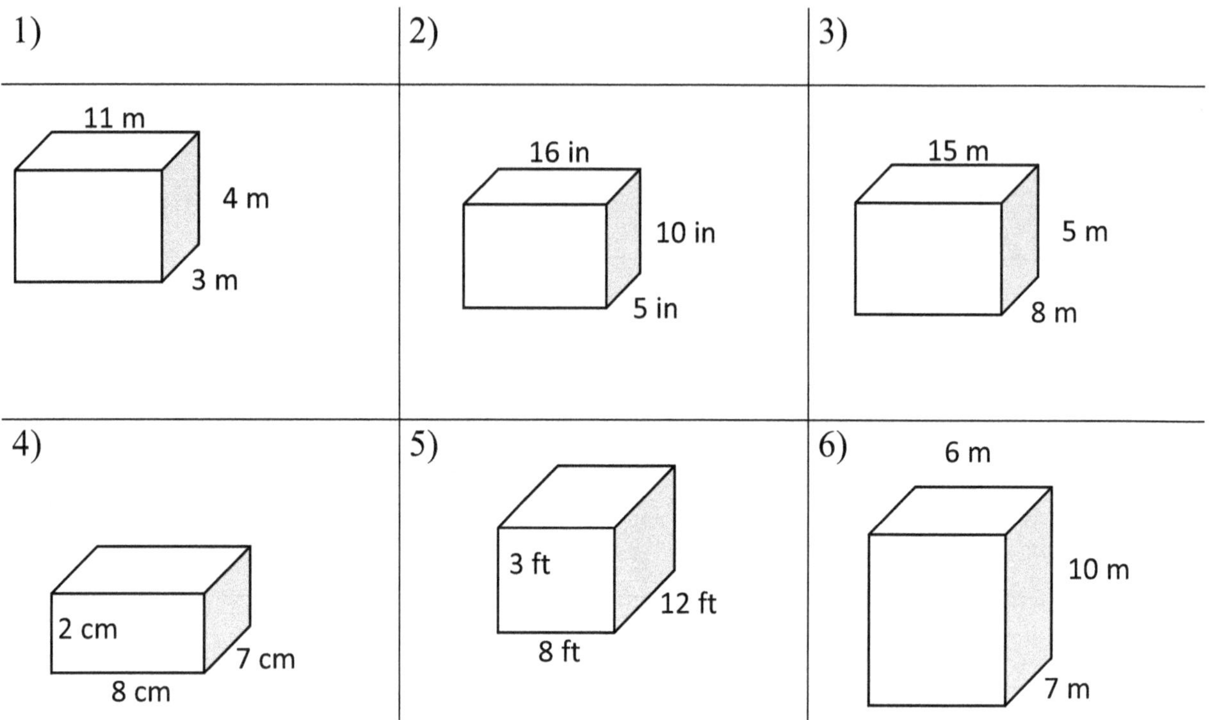

Find the surface area of each Rectangular Prism.

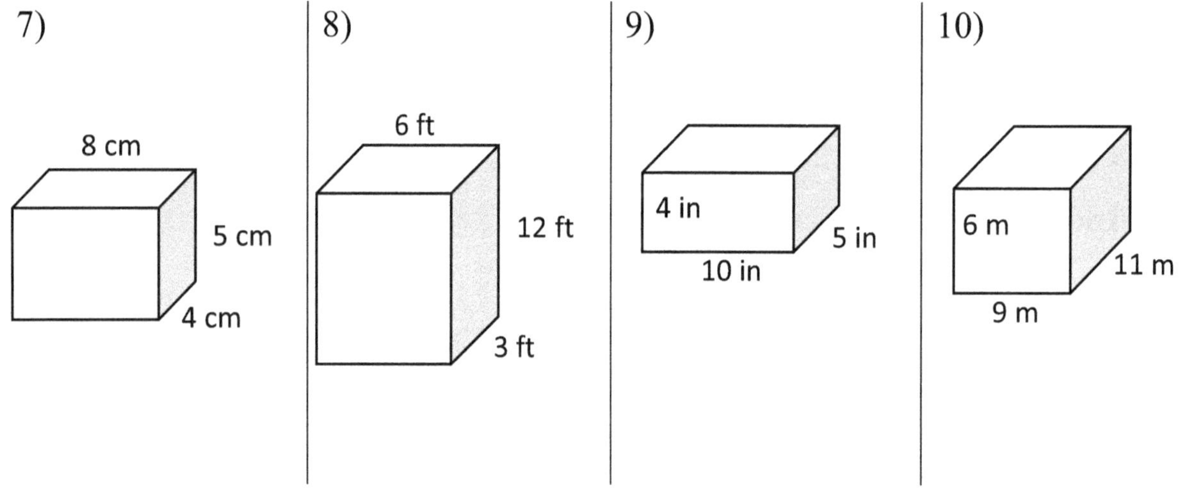

Cylinder

Find the volume of each Cylinder. Round your answer to the nearest tenth. ($\pi = 3.14$)

1)

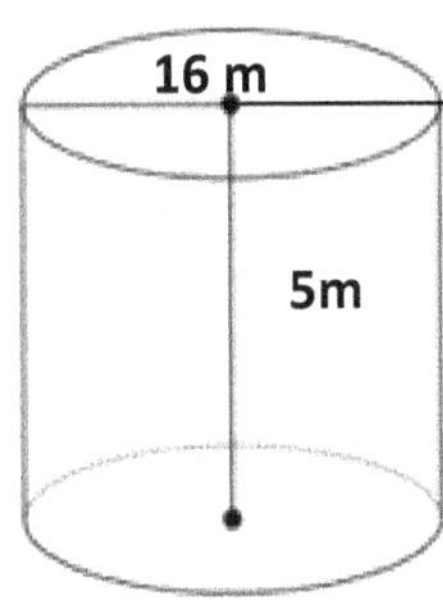

2)

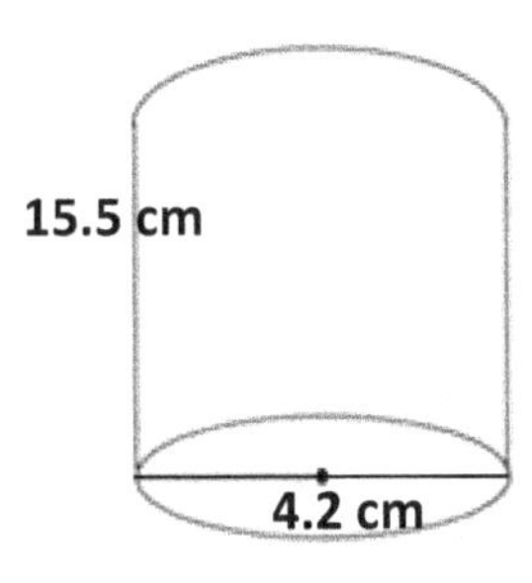

3)

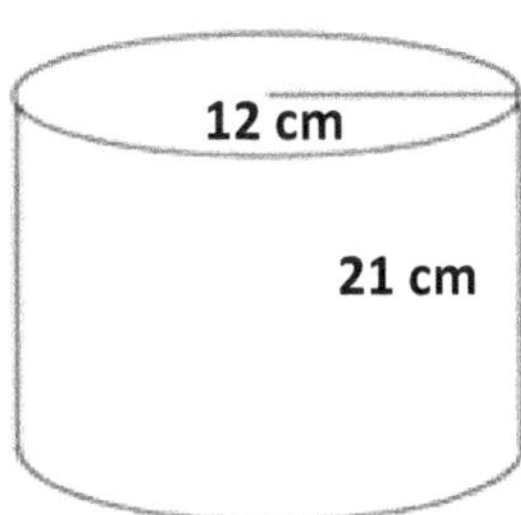

4)

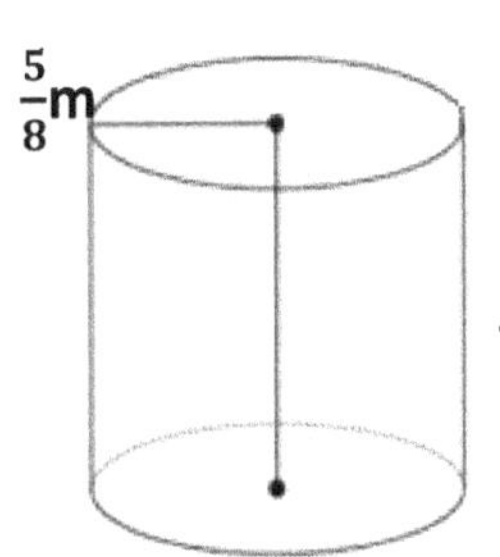

5)

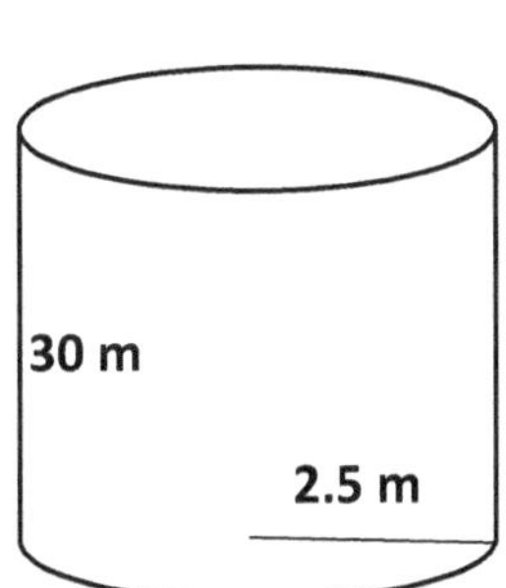

6)

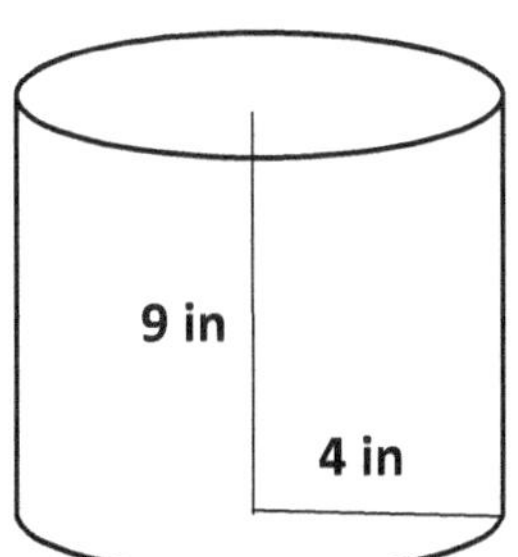

Find the surface area of each Cylinder. ($\pi = 3.14$)

7)

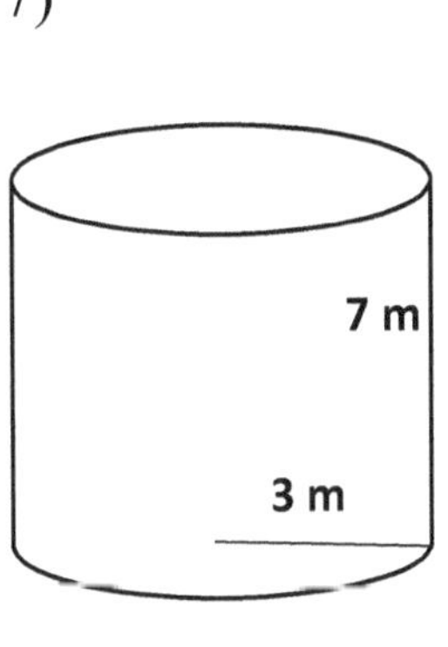

8)

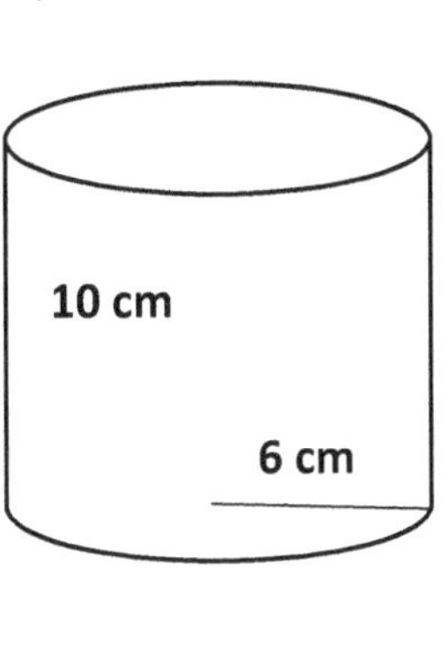

9)

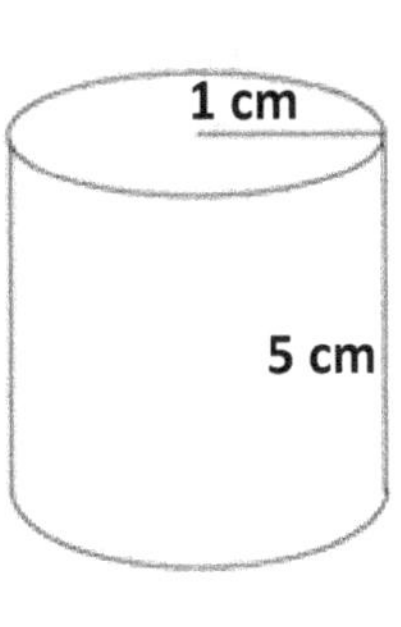

10)

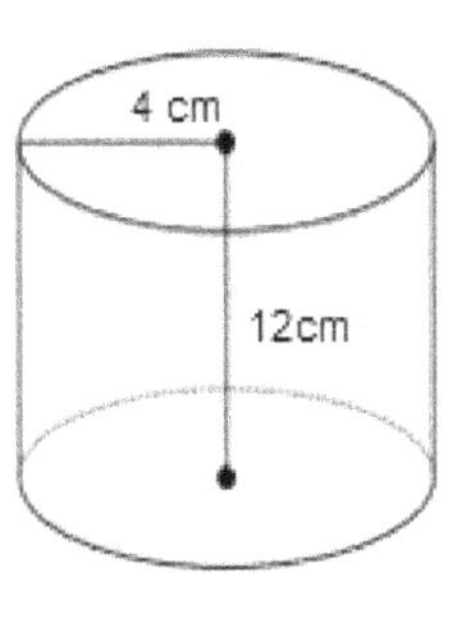

Pyramids and Cone

Find the volume of each Pyramid and Cone. ($\pi = 3.14$)

1)

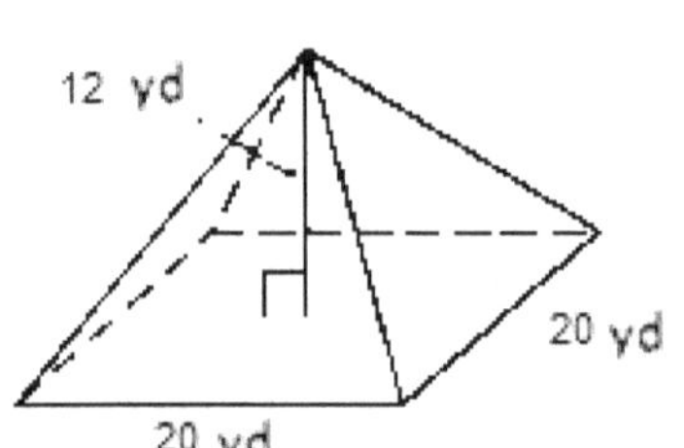

2)

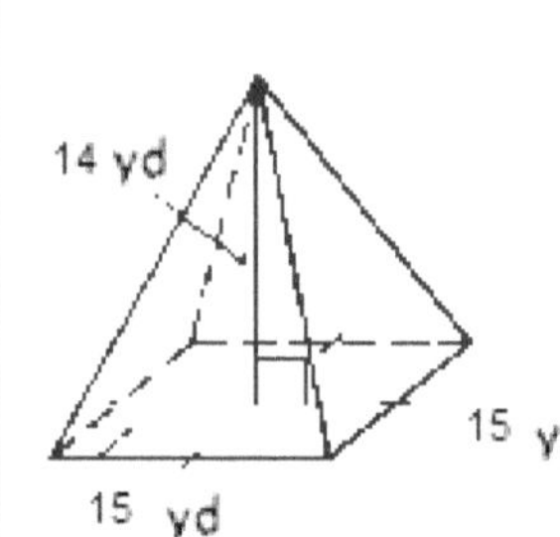

3)

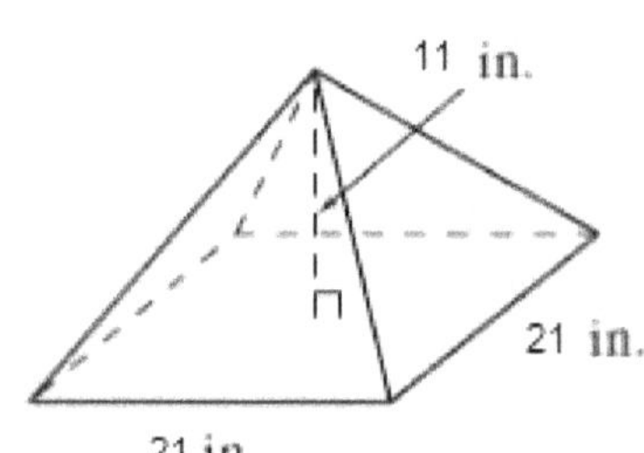

4)

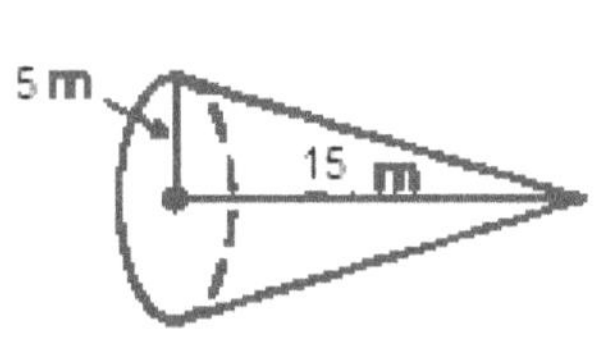

5)

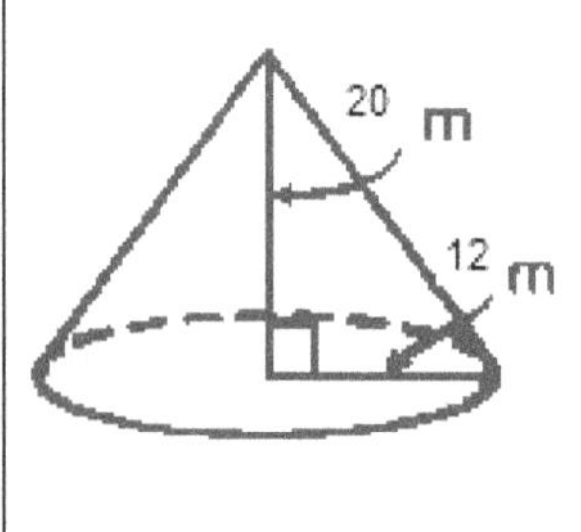

6)

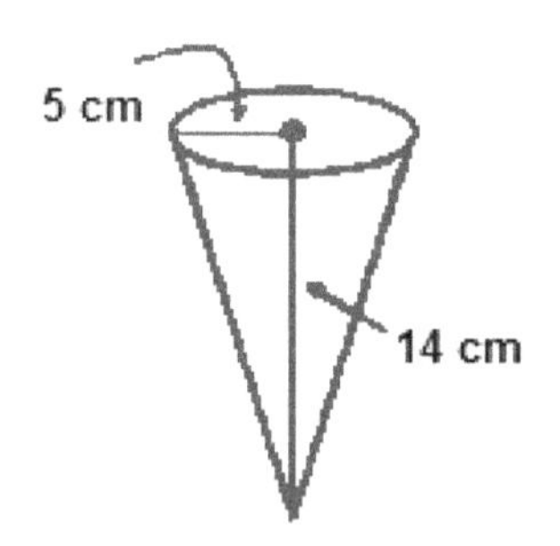

Find the surface area of each Pyramid and Cone. ($\pi = 3.14$)

7)

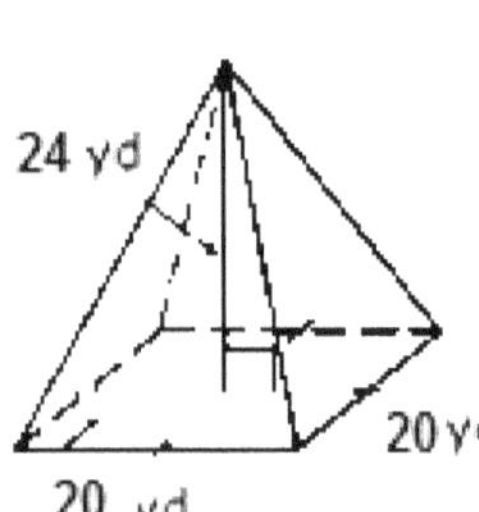

8)

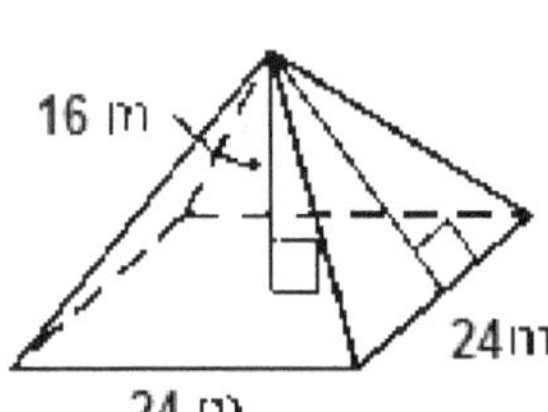

9)

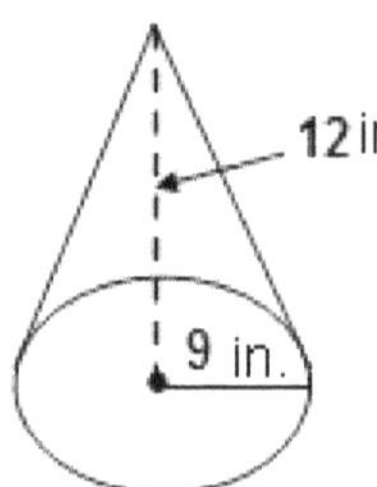

10)

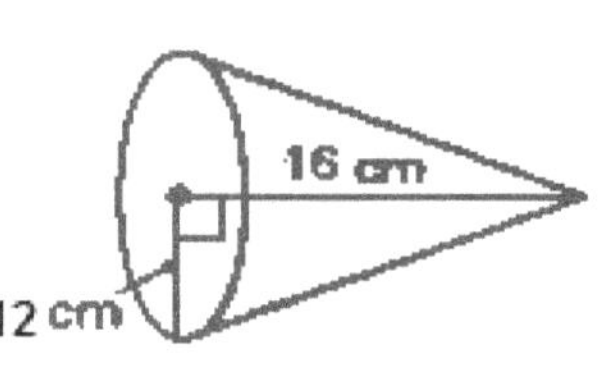

Answers of Worksheets

Angles

1) 16°
2) 96°
3) 59°
4) 34°
5) 70°
6) 52°
7) 90°
8) 75°
9) 33°
10) 75°
11) 70°

Pythagorean Relationship

1) No
2) Yes
3) No
4) Yes
5) Yes
6) No
7) Yes
8) Yes
9) 13
10) 20
11) 17
12) 10
13) 15
14) 30
15) 36
16) 12

Triangles

1) 60°
2) 48°
3) 55°
4) 52°
5) 45°
6) 40°
7) 45°
8) 67°
9) 54 *square unites*
10) 120 *square unites*
11) 90 *square unites*
12) 180 *square unites*

Polygons

1) 52 ft
2) 26 in
3) 112 ft
4) 20 cm
5) 54 m
6) 25 cm
7) 38 in
8) 24 m
9) 30 m^2
10) 300 in^2
11) 160 km^2
12) 49 in^2

Trapezoids

1) 50 cm^2
2) 105 m^2
3) 39 ft^2
4) 60 cm^2
5) 80
6) 24
7) 36
8) 15

Calculate

1) 13 cm
2) 11 ft
3) 12 m
4) 25 ft

Circles

1) 19.63 in^2
2) 78.5 cm^2
3) 254.34 ft^2
4) 12.56 m^2
5) 379.94 cm^2
6) 314 $miles^2$
7) 132.67 in^2
8) 7.07 ft^2
9) 12.56 m^2
10) 113.04 cm^2
11) 38.47 $miles^2$
12) 50.24 ft^2

Circle No.	Radius	Diameter	Circumference	Area
1	1 in	2 in	6.28 in	3.14 in^2
2	5 m	10 m	31.4 m	78.5 m^2
3	3 ft	6 ft	18.84 ft	28.26 ft^2
4	7.5 miles	15 mi	47.1 mi	176.63 mi^2
5	5.5 km	11 km	34.54 km	94.99 km^2
6	7 cm	14 cm	43.96 cm	153.86 cm^2
7	6 ft	12 ft	37.68 feet	113.04 ft^2
8	10 m	20 m	62.8 m	314 m^2
9	9 in	18 in	56.52 in	254.34 in^2
10	4.5 ft	9 ft	28.26 ft	63.585 ft^2

Cubes

1) 12
2) 8 cm^3
3) 216 ft^3
4) 1,331 m^3
5) 2,197 in^3
6) 343 $miles^3$
7) 1.728 km^3
8) 729 cm^3
9) 9.261 ft^3
10) 1,728 mm^3
11) 0.008 in^3
12) 0.001 km^3
13) 27
14) 294 m^2
15) 150 ft^2
16) 121.5 mm^2
17) 7.26 km^2
18) 726 cm^2

Rectangular Prism

1) 132 m^3
2) 800 in^3
3) 600 m^3
4) 112 cm^3
5) 288 ft^3
6) 420 m^3
7) 184 cm^2
8) 252 ft^2
9) 220 in^2
10) 438 m^2

Cylinder

1) 1,004.8 m^3
2) 214.6 cm^3
3) 9,495.4 cm^3
4) 1.1 m^3
5) 588.8 m^3
6) 452.2 in^3
7) 188.4 m^2
8) 602.9 cm^2
9) 37.7 cm^2
10) 401.9 m^2

Pyramids and Cone

1) 1,600 yd^3
2) 1,050 yd^3
3) 1,617 in^3
4) 392.5 m^3
5) 3,014.4 m^3
6) 366.33 cm^3
7) 1,440 yd^2
8) 1,536 m^2
9) 678.24 in^2
10) 1,205.76 cm^2

Chapter 11 :
Statistics and Probability

Topics that you will practice in this chapter:

- ✓ Mean and Median
- ✓ Mode and Range
- ✓ Histograms
- ✓ Stem–and–Leaf Plot
- ✓ Pie Graph
- ✓ Probability Problems
- ✓ Factorials
- ✓ Combinations and Permutation

Mathematics is no more computation than typing is literature.
– John Allen Paulos

Mean and Median

Find Mean and Median of the Given Data.

1) 8, 7, 14, 4, 8

2) 14, 8, 25, 19, 16, 33, 11

3) 23, 18, 15, 12, 17

4) 34, 14, 10, 15, 6, 11

5) 10, 19, 6, 8, 32, 20, 17

6) 17, 26, 39, 69, 20, 6

7) 40, 38, 18, 11, 9, 2, 7, 32,41

8) 24, 21, 31,12,33, 32, 22

9) 16, 14, 20, 41, 15, 20, 38, 4

10)20, 20, 30, 18, 6, 28, 12, 46

11) 12, 7, 10, 11, 16, 22

12) 10, 29, 27, 12, 2, 15, 10, 3

Calculate.

13)In a javelin throw competition, five athletics score 56, 34, 62, 23 and 19 meters. What are their Mean and Median? ________________

14)Eva went to shop and bought 8 apples, 14 peaches, 6 bananas, 4 pineapples and 12 melons. What are the Mean and Median of her purchase? ________________

15)Bob has 17 black pen, 19 red pen, 14 green pens, 20 blue pens and 5 boxes of yellow pens. If the Mean and Median are 19 respectively, what is the number of yellow pens in each box? ________________

Mode and Range

Find Mode and Rage of the Given Data.

1) 4, 3, 7, 3, 3, 4

Mode: _____ Range: _____

2) 18, 18, 24, 26, 18, 8, 14, 22

Mode: _____ Range: _____

3) 8, 8, 8, 16, 19, 22, 20, 9, 13

Mode: _____ Range: _____

4) 24, 24, 14, 28, 20, 18, 20, 24

Mode: _____ Range: _____

5) 6, 21, 27, 24, 27, 27

Mode: _____ Range: _____

6) 21, 8, 8, 7, 8, 12, 10, 22, 18, 13

Mode: _____ Range: _____

7) 7, 4, 4, 6, 13, 13, 13, 0, 2, 2

Mode: _____ Range: _____

8) 5, 8, 5, 14, 12, 14, 3, 5, 18

Mode: _____ Range: _____

9) 7, 7, 7, 12, 7, 3, 8, 16, 3, 17

Mode: _____ Range: _____

10) 15, 15, 19, 16, 4, 16, 10, 15

Mode: _____ Range: _____

11) 6, 6, 5, 6, 42, 13, 19, 2

Mode: _____ Range: _____

12) 8, 8, 9, 8, 9, 4, 34, 22

Mode: _____ Range: _____

Calculate.

13) A stationery sold 12 pencils,56 red pens,24 blue pens,20 notebooks, 12 erasers, 21 rulers and 11 color pencils. What are the Mode and Range for the stationery sells?

Mode: _____ Range: _____

14) In an English test, eight students score 10, 15, 15, 18 18, 16, 15 and 15. What are their Mode and Range? ________________

15) What is the range of the first 6 even numbers greater than 8?

____

Times Series

Use the following Graph to complete the table.

Day	Distance (km)
1	
2	

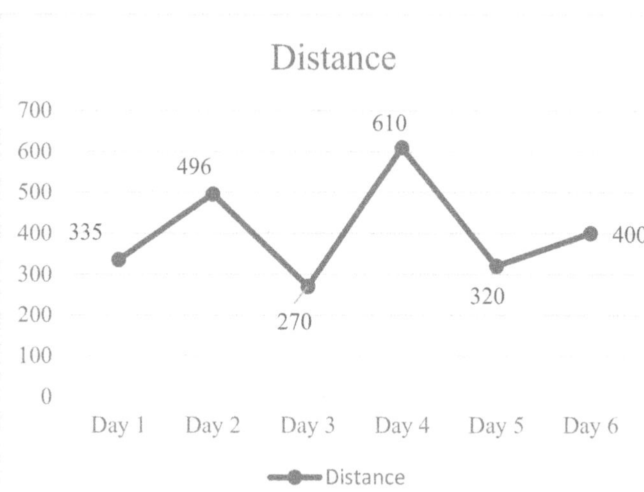

The following table shows the number of births in the US from 2007 to 2012 (in millions).

Year	Number of births (in millions)
2007	4.15
2008	3.70
2009	3.45
2010	3.20
2011	1.75
2012	2.98

Draw a Time Series for the table.

Stem–and–Leaf Plot

Make stem ad leaf plots for the given data.

1) 24, 26,29, 20, 53, 27, 51, 55, 36, 21, 37, 30

Stem	Leaf plot

2) 11, 59, 66, 14, 18, 19, 59, 65, 69, 61, 68, 65

Stem	Leaf plot

3) 121, 55, 66, 54, 112, 128, 63, 125, 59, 123, 68, 119

Stem	Leaf plot

4) 51, 32, 100, 56, 84, 36, 107, 56, 85, 39, 56, 106, 89

Stem	Leaf plot

5) 33, 89, 19, 87, 81, 16, 11, 30, 86, 35, 17, 35, 13

Stem	Leaf plot

6) 60, 92, 22, 25, 67, 93, 95, 62, 21, 64, 98, 29

Stem	Leaf plot

Pie Graph

The circle graph below shows all Robert's expenses for last month. Robert spent $140 on his hobbies last month.

Answer following questions based on the Pie graph.

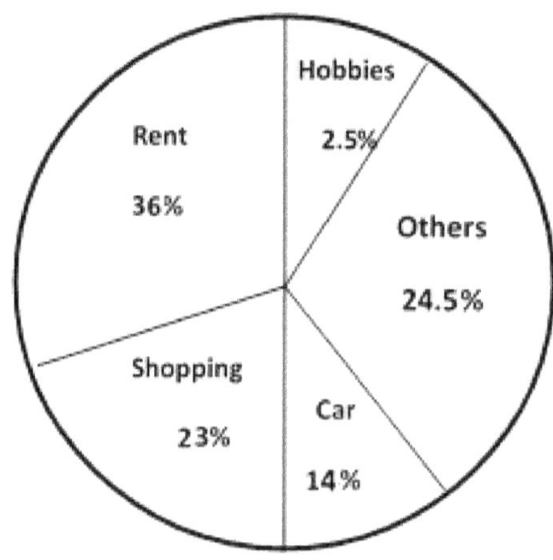

1) How much was Robert's total expenses last month? ________________

2) How much did Robert spend on his car last month? ________________

3) How much did Robert spend for shopping last month? __________

4) How much did Robert spend on his rent last month? ________________

5) What fraction is Robert's expenses for his rent and car out of his total expenses last month? ________________

Probability Problems

Calculate.

1) A number is chosen at random from 1 to 10. Find the probability of selecting number 6 or smaller numbers. ____________________

2) Bag A contains 18 red marbles and 6 green marbles. Bag B contains 16 black marbles and 8 orange marbles. What is the probability of selecting a green marble at random from bag A? What is the probability of selecting a black marble at random from Bag B? ____________________

3) A number is chosen at random from 1 to 20. What is the probability of selecting multiples of 4? ____________________

4) A card is chosen from a well-shuffled deck of 52 cards. What is the probability that the card will be a queen? ____________________

5) A number is chosen at random from 1 to 15. What is the probability of selecting a multiple of 3 or 5? ____________________

A spinner numbered 1–8, is spun once. What is the probability of spinning ...?

6) an Odd number? __________ 7) a multiple of 2? _____

8) a multiple of 5? _____ 9) number 10? ________

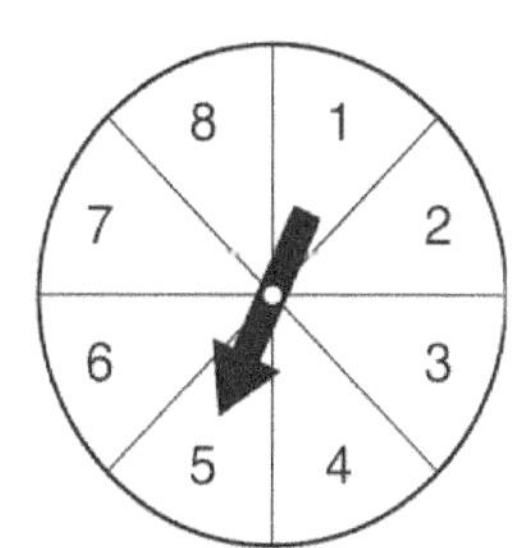

Factorials

Determine the value for each expression.

1) $4!\ +0!\ =$

2) $2!+5!\ =$

3) $(2!)^2 =$

4) $5!-3! =$

5) $6!-3!+10 =$

6) $3!\times 4-15 =$

7) $(2!+3!)^2 =$

8) $(4!-3!)^2 =$

9) $(3!\,0!)^2-10 =$

10) $\frac{10!}{8!} =$

11) $\frac{6!}{4!} =$

12) $\frac{6!}{5!} =$

13) $\frac{15!}{13!} =$

14) $\frac{n!}{(n-3)!} =$

15) $\frac{(n+2)!}{n!} =$

16) $\frac{(2+2!)^3}{2!} =$

17) $\frac{5(n+2)!}{(n+1)!} =$

18) $\frac{22!}{20!4!} =$

19) $\frac{13!}{11!3!} =$

20) $\frac{9\times 210!}{3(7\times 30)!} =$

21) $\frac{32!}{31!2!} =$

22) $\frac{11!12!}{10!13!} =$

23) $\frac{16!15!}{14!14!} =$

24) $\frac{(5\times 3)!}{0!14!} =$

25) $\frac{4!(5n-2)!}{(5n)!} =$

26) $\frac{4n(4n+7)!}{(4n+8)!} =$

27) $\frac{(n-2)!(n+1)}{(n+2)!} =$

Combinations and Permutations

Calculate the value of each.

1) $6! =$ ____
2) $2! \times 5! =$ ____
3) $3 \times 4! =$ ____
4) $5! + 3! =$ ____
5) $7! =$ ____
6) $4! =$ ____
7) $3! + 3! =$ ____
8) $7! - 5! =$ ____

Find the answer for each word problems.

9) Susan is baking cookies. She uses sugar, butter, Vanilla, eggs and flour. How many different orders of ingredients can she try? ____________

10) Albert is planning for his vacation. He wants to go to museum, watch a movie, go to the beach, play the game and play football. How many ways of ordering are there for him? ____________

11) How many 4-digit numbers can be named using the digits 3, 4, 5, and 6 without repetition? ____________

12) In how many ways can 5 boys be arranged in a straight line? ____________

13) In how many ways can 6 athletes be arranged in a straight line? ____________

14) A professor is going to arrange her 7 students in a straight line. In how many ways can she do this? ____________

15) How many code symbols can be formed with the letters for the word GAMES? ____________

16) In how many ways a team of 7 basketball players can choose a captain and co-captain? ____________

Answers of Worksheets

Mean and Median

1) Mean: 8.2, Median: 8
2) Mean: 18, Median: 16
3) Mean: 17, Median: 17
4) Mean: 15, Median: 12.5
5) Mean: 16, Median: 17
6) Mean: 29.5, Median: 23
7) Mean: 22, Median: 18
8) Mean: 25, Median: 24
9) Mean: 21, Median: 18
10) Mean: 22.5, Median: 20
11) Mean: 13, Median: 11.5
12) Mean: 13.5, Median: 11
13) Mean: 38.8, Median: 34
14) Mean: 8.8, Median: 8
15) 5

Mode and Range

1) Mode: 3, Range: 4
2) Mode: 18, Range: 18
3) Mode: 8, Range: 14
4) Mode: 24, Range: 14
5) Mode: 27, Range: 21
6) Mode: 8, Range: 15
7) Mode: 13, Range: 13
8) Mode: 5, Range: 15
9) Mode: 7, Range: 14
10) Mode: 15, Range: 15
11) Mode: 6, Range: 40
12) Mode: 8, Range: 30
13) Mode: 12, Range: 45
14) Mode: 15, Range: 8
15) 10

Time series

Day	Distance (km)
1	335
2	496
3	270
4	610
5	320
6	400

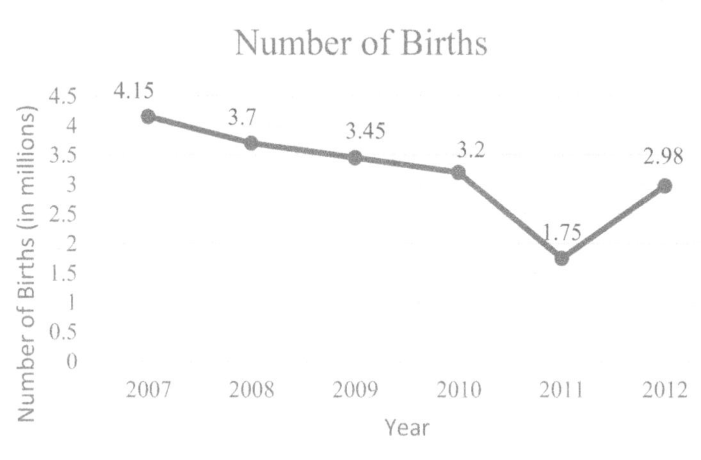

Stem–And–Leaf Plot

1)

Stem	leaf
2	0 1 4 6 7 9
3	0 6 7
5	1 3 5

2)

Stem	leaf
1	1 4 8 9
5	9 9
6	1 5 5 6 8 9

3)

Stem	leaf
5	4 5 9
6	3 6 8
11	2 9
12	1 3 5 8

4)

Stem	leaf
3	2 6 9
5	1 6 6 6
8	4 5 9
10	0 6 7

5)

Stem	leaf
1	1 3 6 7 9
3	0 3 5 5
8	1 6 7 9

6)

Stem	leaf
2	2 1 5 9
6	0 2 4 7
9	2 3 5 8

Pie Graph

1) $5,600
2) $784
3) $1,288
4) $2,016
5) $\frac{1}{2}$

Probability Problems

1) $\frac{3}{5}$
2) $\frac{1}{4}, \frac{2}{3}$
3) $\frac{1}{4}$
4) $\frac{1}{13}$
5) $\frac{7}{15}$
6) $\frac{1}{2}$
7) $\frac{1}{2}$
8) $\frac{1}{8}$
9) 0

Factorials

1) 25
2) 122
3) 4
4) 114
5) 724
6) 9
7) 64
8) 324
9) 26
10) 90
11) 30
12) 6
13) 210
14) $n(n-1)(n-2)$
15) $(n+1)(n+2)$
16) 32
17) $5(n+2)$
18) 19.25
19) 26
20) 3
21) 16
22) $\frac{11}{13}$
23) 3,600
24) 15
25) $\frac{24}{5n(5n-1)}$
26) $\frac{n}{(n+2)}$
27) $\frac{1}{n(n-1)(n+2)}$

Combinations and Permutations

1) 720
2) 240
3) 72
4) 126
5) 5,040
6) 24
7) 12
8) 4,920
9) 120
10) 120
11) 24
12) 120
13) 720
14) 5,040
15) 120
16) 42

Chapter 12 : MCAS Math Practice Tests

Time to Test

Time to refine your skill with a practice examination.

Take a REAL MCAS Mathematics test to simulate the test day experience. After you've finished, score your test using the answer key.

Before You Start

- You'll need a pencil, calculator, and a timer to take the test.
- It's okay to guess. You won't lose any points if you're wrong.
- After you've finished the test, review the answer key to see where you went wrong.

Graphing calculators are permitted for MCAS Tests Grade 8

Good Luck!

Massachusetts Comprehensive Assessment System
Grade 8 Mathematics Reference Sheet

PERIMETER FORMULAS

square. . . . . $P = 4s$

rectangle. . . . . $P = 2b + 2h$
OR
$P = 2l + 2w$

triangle. . . . . $P = a + b + c$

AREA FORMULAS

square. . . . . $A = s^2$

rectangle. . . . . $A = bh$
OR
$A = lw$

parallelogram . . . . . $A = bh$

triangle. . . . . $A = \frac{1}{2}bh$

trapezoid. . . . . $A = \frac{1}{2}h(b_1 + b_2)$

circle. . . . . $A = \pi r^2$

TOTAL SURFACE AREA FORMULAS

rectangular prism . . $SA = 2(lw) + 2(hw) + 2(lh)$

cylinder . . . . $SA = 2\pi r^2 + 2\pi rh$

sphere. . . . . $SA = 4\pi r^2$

VOLUME FORMULAS

rectangular prism . . . . . . $V = lwh$
OR
$V = Bh$
(B = area of a base)

cube. . . . . $V = s^3$
(s = length of an edge)

cylinder . . . . $V = \pi r^2 h$

sphere . . . . $V = \frac{4}{3}\pi r^3$

CIRCLE FORMULAS

$C = 2\pi r$
OR
$C = \pi d$

$A = \pi r^2$

PYTHAGOREAN THEOREM

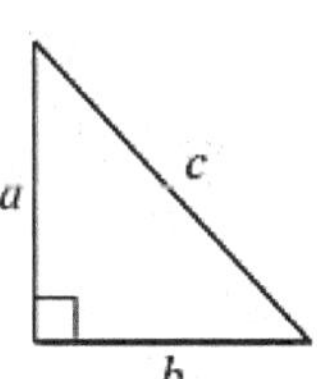

$a^2 + b^2 = c^2$

Massachusetts Comprehensive Assessment System Practice Test 1

Mathematics

GRADE 8

Administered *Month Year*

Session 1

- ❖ **Graphing Calculators are permitted for this practice test.**
- ❖ **Time for Session 1: 85 Minutes**

1) When a number is subtracted from 50 and the difference is divided by that number, the result is 4. What is the value of the number?

A. 8

B. 10

C. 14

D. 12

2) How much interest is earned on a principal of $15,000 invested at an interest rate of 3.15% for four years?

A. $9,180

B. $18,800

C. $1,820

D. $1,890

3) A swimming pool holds 1,820 cubic feet of water. The swimming pool is 26 feet long and 14 feet wide. How deep is the swimming pool?

Write your answer in the box below.

4) In two successive years, the population of a town is increased by 12% and 35%. What percent of the population is increased after two years?

A. 25.2%

B. 51.2%

C. 252%

D. 512 %

5) Which graph shows a non–proportional linear relationship between x and y?

A.

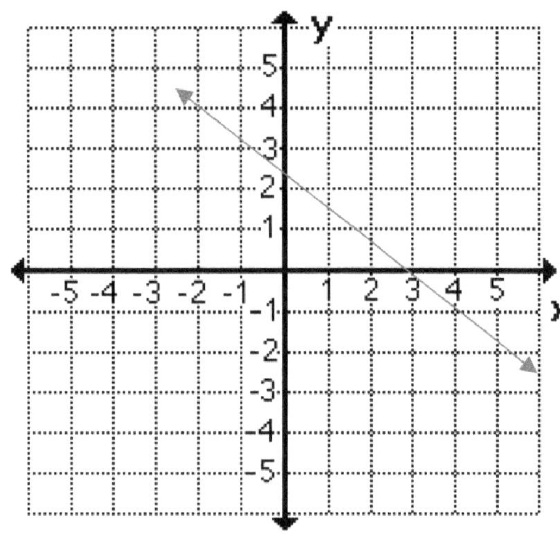

B.

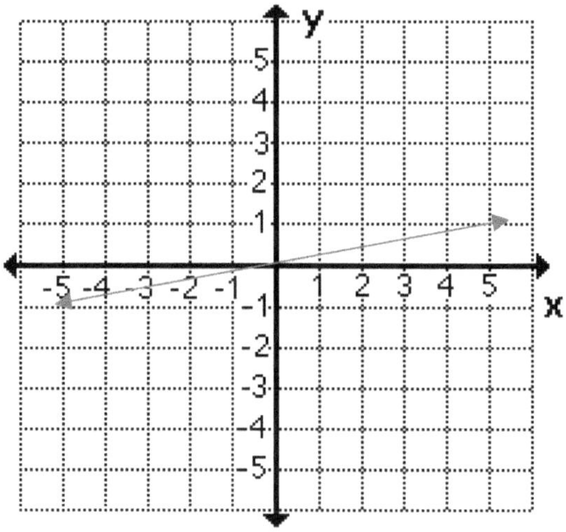

C.

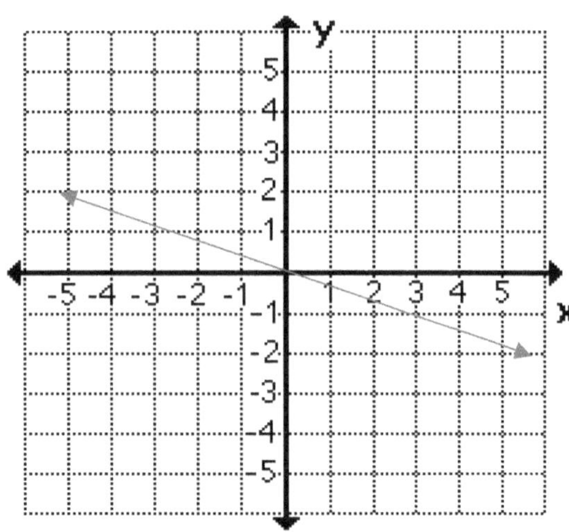

D.

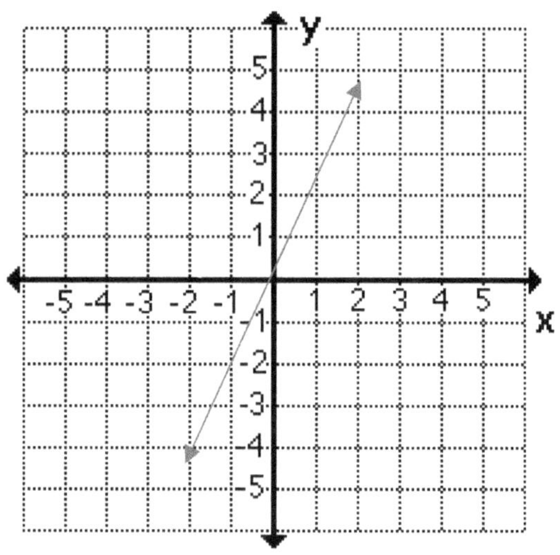

6) Five years ago, Amy was four times as old as Mike was. If Mike is 9 years old now, how old is Amy?

A. 20

B. 21

C. 28

D. 30

7) What is the value of $|-47 + 22| - |-7(-8)|$?

A. -13

B. $+14$

C. -25

D. -31

8) What is the solution of the following system of equations?

$$\begin{cases} \frac{x}{6} + \frac{y}{3} = 2 \\ \frac{-2y}{3} - x = -8 \end{cases}$$

A. $x = 6, y = 3$

B. $x = 6, y = 4$

C. $x = 6, y = -3$

D. $x = -6, y = -4$

9) If a gas tank can hold 64 gallons, how many gallons does it contain when it is $\frac{3}{8}$ full?

A. 24

B. 28

C. 32

D. 46

10) Jack earns $360 for his first 45 hours of work in a week and is then paid 3 times his regular hourly rate for any additional hours. This week, Jack needs $960 to pay his rent, bills and other expenses. How many hours must he work to make enough money in this week?

Write your answer in the box below.

Session 2

- ❖ **Graphing Calculators are permitted for this practice test.**
- ❖ **Time for Session 2: 85 Minutes**

11) The sum of six different negative integers is -45. If the smallest of these integers is -10, what is the largest possible value of one of the six four integers?

A. -4

B. -7

C. -8

D. -5

12) In the rectangle below if $y > 3$ cm and the area of rectangle is 16 cm^2 and the perimeter of the rectangle is 20 cm, what is the value of x and y respectively?

A. 4, 5

B. 4, 4

C. 2, 8

D. 3, 8

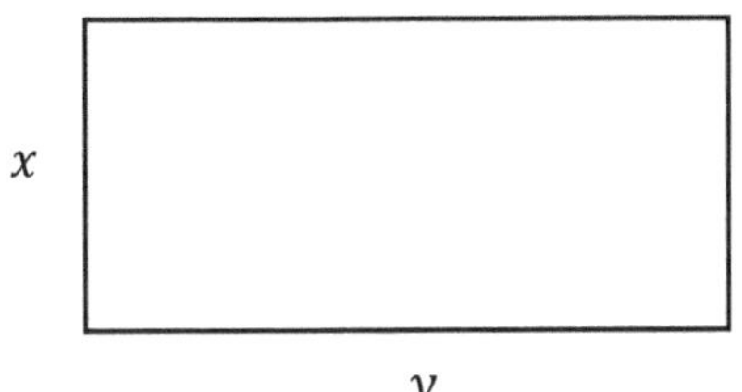

13) The following table represents the value of x and function $f(x)$. Which of the following could be the equation of the function $f(x)$?

A. $f(x) = 2x^2 + 1$

B. $f(x) = 3x + 1$

C. $f(x) = \sqrt{4x - 1}$

D. $f(x) = 4\sqrt{x} + 1$

x	$f(x)$
1	5
4	9
9	13
16	17

Questions 14, 15 are based on the following data

Types of air pollutions in 10 cities of a country

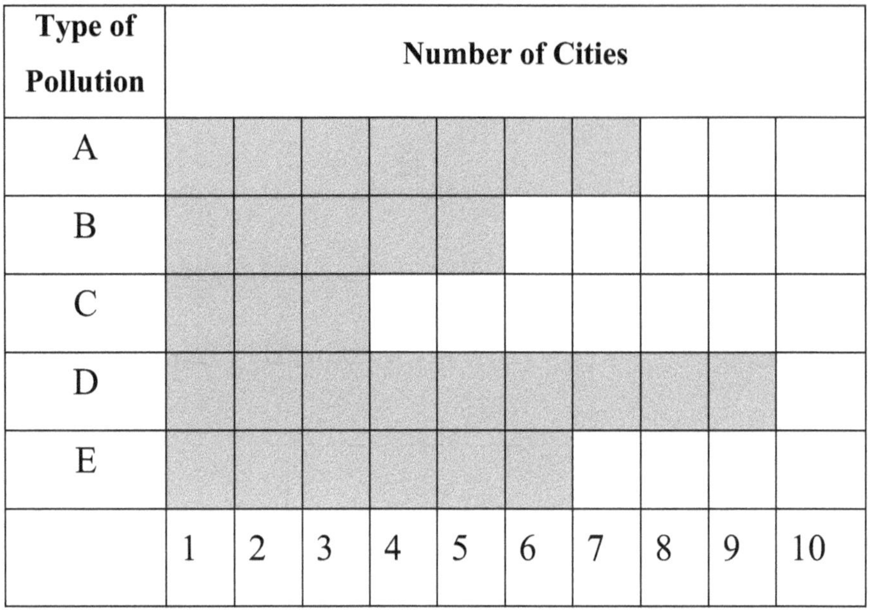

Type of Pollution	Number of Cities									
A	■	■	■	■	■	■	■			
B	■	■	■	■	■					
C	■	■	■							
D	■	■	■	■	■	■	■	■	■	
E	■	■	■	■	■	■				
	1	2	3	4	5	6	7	8	9	10

14) If a is the mean (average) of the number of cities in each pollution type category, b is the mode, and c is the median of the number of cities in each pollution type category, then which of the following must be true?

A. $b < c < a$

B. $a < c < b$

C. $a = c$

D. $c < a = b$

15) How many cities should be added to type of pollutions C until the ratio of cities in type of pollution C to cities in type of pollution E will be 1.5?

A. 4

B. 6

C. 7

D. 9

16) What is the area of the shaded region if the diameter of the bigger circle is 8 inches and the diameter of the smaller circle is 6 inches?

A. 2π

B. 7π

C. 11π

D. 28π

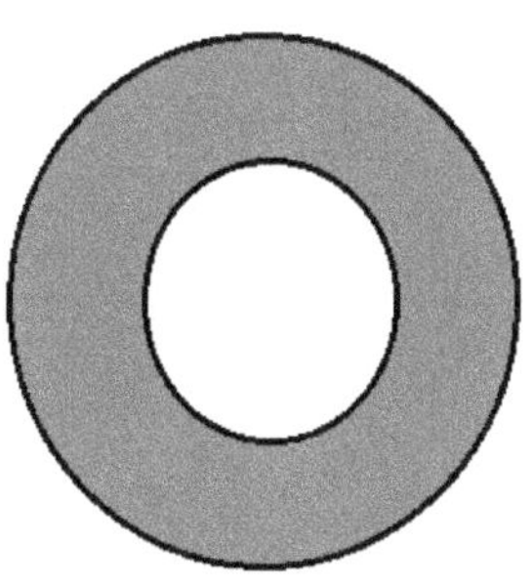

17) What is the sum of $\sqrt{4x+9}$ and $\sqrt{4x}-2$ when $\sqrt{x}=2$?

Write your answer in the box below.

18) In the following figure, point M lies on the line A, what is the value of y if $x=21$?

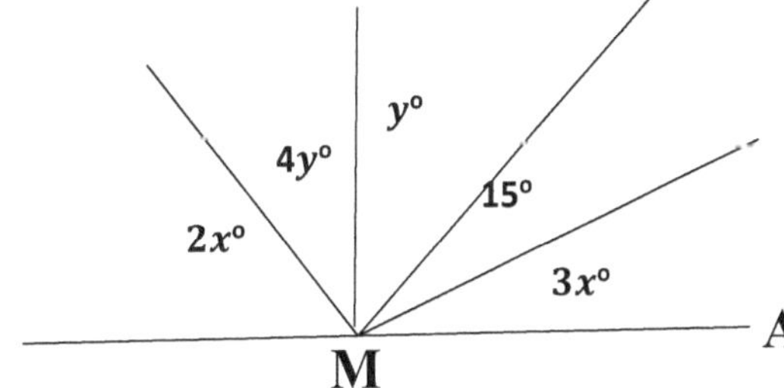

Write your answer in the box below.

19) In the following right triangle, if the sides AB and AC become triple longer, what will be the ratio of the perimeter of the triangle to its area?

A. 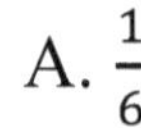

B. 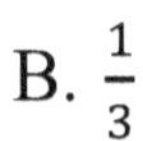

C. 6

D. 3

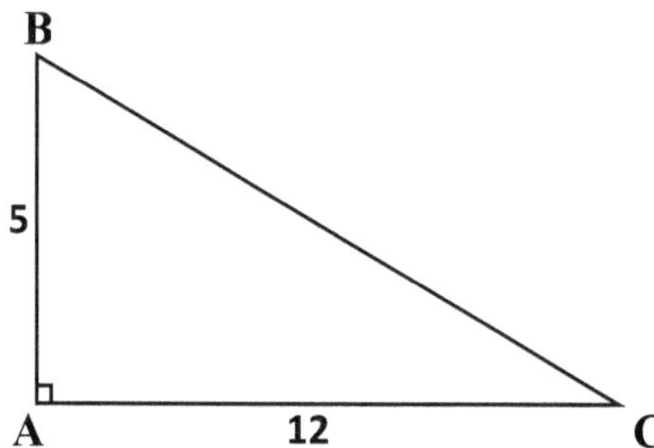

20) In the xy-plane, the point $(3, 4)$ and $(8,9)$ are on the line A. Which of the following equations of lines is parallel to line A?

A. $y - x = 8$

B. $y = \frac{x}{4} + \frac{3}{8}$

C. $y = 3x + 8$

D. $3y - 8x = 1$

Practice Test 1

This is the End of this Section.

Massachusetts Comprehensive

Assessment System Practice Test 2

Mathematics

GRADE 8

Administered *Month Year*

Session 1

- Graphing Calculators are permitted for this practice test.
- Time for Session 1: 85 Minutes

1) A tree 18 feet tall casts a shadow 66 feet long. Jack is 3 feet tall. How long is Jack's shadow?

A. 33 ft
C. 11 ft
B. 44 ft
D. 22 ft

2) If $(2^a)^b = 512$, then what is the value of $a \times b$?

A. $6a$
C. 9
B. $9b$
D. 6

3) A bag contains 22 balls: seven green, six black, three blue, five red and one white. If 16 balls are removed from the bag at random, what is the probability that a white ball has been removed?

A. $\frac{8}{11}$
C. $\frac{1}{22}$
B. $\frac{1}{8}$
D. $\frac{1}{16}$

4) What is the x-intercept of the line with equation$-7x + 6y = -42$?

A. -6
C. $+7$
B. $+6$
D. $-\frac{6}{7}$

5) Which of the following expressions is equivalent to $7xy(2x - 4y)$?

A. $28yx^2 - 14xy$
C. $-28xy^2 + 14x^2y$
B. $28x^2 - 14y^2$
D. $28xy^2 - 14yx^2$

6) The equation of a line is given as: $y = -7x + 2$. Which of the following points does not lie on the line?

A. $(-1, 9)$

C. $(-3, 23)$

B. $(2, -12)$

D. $(2, -12)$

7) Which of the following is equivalent to $-41 < -6x + 1 < 13$?

A. $-2 < x < 2$

B. $-2 < x < 7$

C. $-7 < x < 2$

D. $-7 < x < -7$

8) The perimeter of the trapezoid below is 51 cm. What is its area?

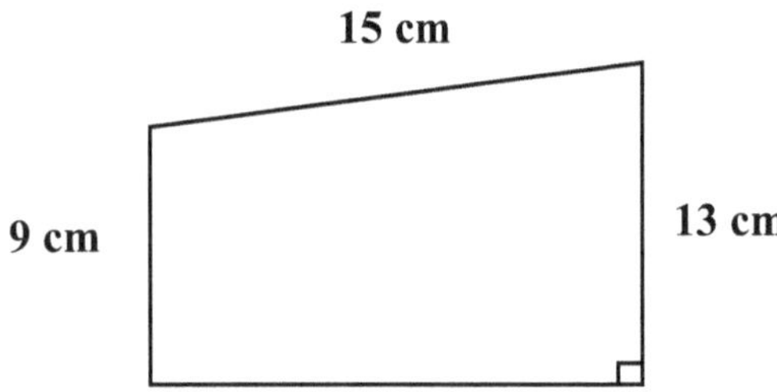

Write your answer in the box below.

9) Which graph does not represent y as a function of x?

A.

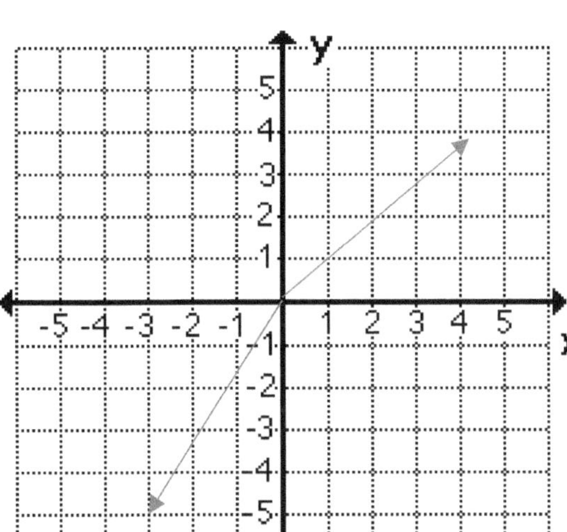

B.

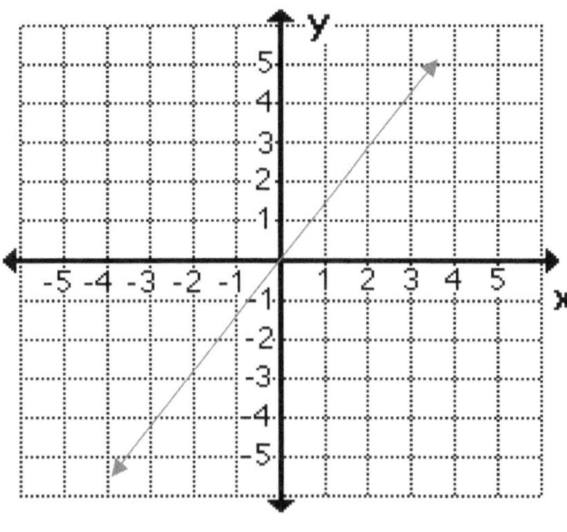

C

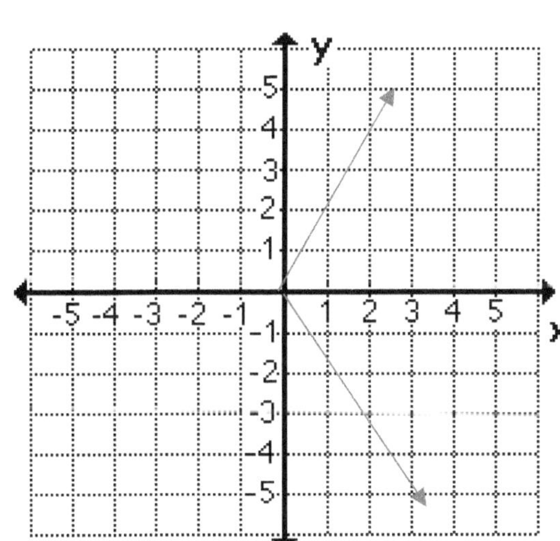

10) Three eighth of 24 is equal to $\frac{3}{7}$ of what number?

Write your answer in the box below.

Session 2

- Graphing Calculators are permitted for this practice test.
- Time for Session 2: 85 Minutes

11) The marked price of a computer is E Euro. Its price decreased by 42% in March and later increased by 15 % in April. What is the final price of the computer in E Euro?

A. 0.076 E

B. 0.776E

C. 0.667 E

D. 0.07 E

12) Triangle ABC is graphed on a coordinate grid with vertices at A $(1,-4)$, B $(-3,5)$ and C $(-1,-8)$. Triangle ABC is reflected over x axes to create triangle A' B' C'. Which order pair represents the coordinate of B'?

A. $(-3,5)$

B. $(3,-5)$

C. $(-3,-5)$

D. $(-5,3)$

13) What is the volume of the following triangular prism?

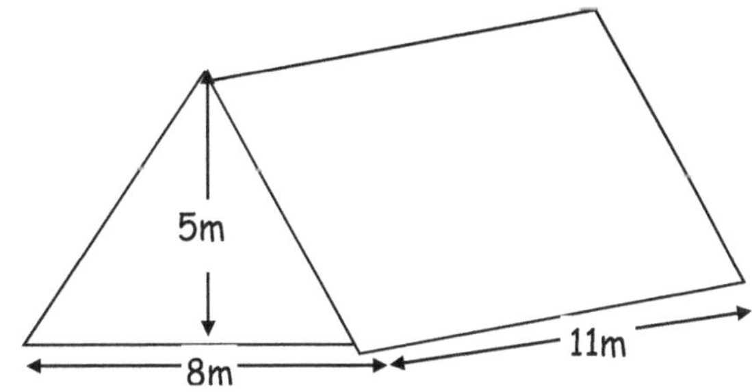

Write your answer in the box below.

14) Which of the following point is the solution of the system of equations?

$$\begin{cases} 3x - 2y = -3 \\ 7x - 5y = -6 \end{cases}$$

A. $(3, 3)$

B. $(-3, -3)$

C. $(-3, 3)$

D. $(3, -3)$

15) What is the distance between the points $(8, -4)$ and $(4, -1)$?

A. 6

B. 8

C. 15

D. 5

16) Which of the following lines is parallel to the graph of $y = 9x - 4$?

A. $9x - 4y = 1$

B. $9x - 7y = 0$

C. $9x - y = -7$

D. $4x + 2y = 9$

17) The average weight of 13 girls in a class is 35 kg and the average weight of 22 boys in the same class is 45 kg. What is the average weight of all the 35 students in that class?

A. 43.15 Kg

B. 42.61 Kg

C. 41.29 Kg

D. 38.56 Kg

18) Which of the following equations has a graph that is a straight line?

A. $y = x^3 - 7$

B. $y^4 - x^4 = 12$

C. $8x - 5y = 13$

D. $3x^2 + xy = -5x$

19) An angle is equal to one eighth of its supplement. What is the measure of that angle?

Write your answer in the box below.

20) A card is drawn at random from a standard 52–card deck, what is the probability that the card is of hearts or diamonds or spades? (The deck includes 13 of each suit clubs, diamonds, hearts, and spades)

A. $\frac{3}{13}$

B. $\frac{3}{4}$

C. $\frac{1}{4}$

D. $\frac{1}{3}$

Practice Test 2

This is the End of this Section.

Chapter 13 : Answers and Explanations

MCAS Practice Tests

Answer Key

❋ Now, it's time to review your results to see where you went wrong and what areas you need to improve!

MCAS - Mathematics

Practice Test - 1			
1	B	**11**	D
2	D	**12**	C
3	5	**13**	D
4	B	**14**	C
5	A	**15**	B
6	B	**16**	B
7	D	**17**	7
8	A	**18**	12
9	A	**19**	B
10	70	**20**	A

Practice Test - 2			
1	C	**11**	C
2	C	**12**	C
3	C	**13**	220
4	B	**14**	B
5	C	**15**	D
6	D	**16**	C
7	B	**17**	C
8	154	**18**	C
9	C	**19**	20
10	21	**20**	B

Practice Test 1
Answers and Explanations

1) Answer: B

Let x be the number. Write the equation and solve for x.

$\frac{(50-x)}{x} = 4$ (cross multiply)

$(50 - x) = 4x$, then add x both sides. $50 = 5x$,

now divide both sides by 10. $\Rightarrow x = 10$.

2) Answer: D

Use simple interest formula: $I = prt$ (I = interest, p = principal, r = rate, t = time)

$I = (15{,}000)(0.0315)(4) = 1{,}890$

3) Answer: 5

Use formula of rectangle prism volume.

V = (length) (width) (height) ⇒ 1,820 = (26) (14) (height)

⇒ height =1,820 ÷ 364 = 5

4) Answer: B

The population is increased by 12% and 35%. 12% increase changes the population to 112% of original population.

For the second increase, multiply the result by 135%.

(1.12) × (1.35) = (1.512) =151.2 %

51.2 percent of the population is increased after two years.

5) Answer: A

A linear equation is a relationship between two variables, x and y, that can be put in the form $y = mx + b$.

A non-proportional linear relationship takes on the form $y = mx + b$, where b ≠ 0 and its graph is a line that does not cross through the origin.

6) Answer: B

Five years ago, Amy was four times as old as Mike. Mike is 9 years now. Therefore, 5 years ago Mike was 4 years.

Five years ago, Amy was: $A = 4 \times 4 = 16$

Now Amy is 21 years old: 16 + 5 = 21

7) Answer: D

$|-47 + 22| - |-7(-8)| = |-25| - |56| = 25 - 56 = -31$

8) Answer: A

$\begin{cases} \frac{x}{6} + \frac{y}{3} = 2 \\ \frac{-2y}{3} - x = -8 \end{cases}$ → Multiply the top equation by 6. Then,

$\begin{cases} x + 2y = 12 \\ \frac{-2y}{3} - x = -8 \end{cases}$ → Add two equations.

$\frac{4}{3}y = 4 \rightarrow y = 3$, plug in the value of y into the first equation→ $x = 6$

9) Answer: A

$\frac{3}{8} \times 64 = \frac{192}{8} = 24$

10) Answer: 70

The amount of money that jack earns for one hour: $\frac{\$360}{45} = \$8$

Number of additional hours that he works to make enough money is: $\frac{\$960-\$360}{3\times\$\,8} = 25$

Number of total hours is: $45 + 25 = 70$

11) Answer: D

The smallest number is -10. To find the largest possible value of one of the other six integers, we need to choose the smallest possible integers for five of them. Let x be the largest number. Then:

$-45 = (-10) + (-9) + (-8) + (-7) + (-6) + x \rightarrow -45 = -40 + x$

$\rightarrow x = -45 + 40 = -5$

12) Answer: C

The perimeter of the rectangle is: $2x + 2y = 20 \rightarrow x + y = 10 \rightarrow x = 10 - y$

The area of the rectangle is: $x \times y = 16 \rightarrow (10 - y)(y) = 16$

$\rightarrow y^2 - 10y + 16 = 0$

Solve the quadratic equation by factoring method.

$(y-2)(y-8)=0 \rightarrow y=2$ (Unacceptable, because y must be greater than 3) or $y=8$; If $y=8 \rightarrow x=10-y \rightarrow x=10-8 \rightarrow x=2$

13) Answer: D

A. $f(x)=2x^2+1$ if $x=4 \rightarrow f(4)=2(4)^2+1=33 \neq 9$

B. $f(x)=3x+1$ if $x=4 \rightarrow f(4)=3(4)+1=13 \neq 9$

C. $f(x)=\sqrt{4x-1}$ if $x=4 \rightarrow f(4)=\sqrt{4(4)-1}=\sqrt{15} \neq 9$

D. $f(x)=4\sqrt{x}+1$ if $x=4 \rightarrow f(4)=4\sqrt{4}+1=9$

14) Answer: C

Let's find the mean (average), mode and median of the number of cities for each type of pollution.

Number of cities for each type of pollution: 7, 5, 3, 9, 6

$average\ (mean)=\frac{sum\ of\ terms}{number\ of\ terms}=\frac{7+5+3+9+6}{5}=\frac{30}{5}=6$

Median is the number in the middle. To find median, first list numbers in order from smallest to largest: 3, 5, 6, 7, 9

Median of the data is 6.

Mode is the number which appears most often in a set of numbers. Therefore, there is no mode in the set of numbers.

Median = Mean, then, $a=c$

15) Answer: B

Let the number of cities should be added to type of pollutions C be x. Then:

$\frac{x+3}{6}=1.5 \rightarrow x+3=6\times 1.5 \rightarrow x+3=9 \rightarrow x=6$

16) Answer: B

To find the area of the shaded region subtract smaller circle from bigger circle.

S bigger – S smaller = π (r bigger)2 – π (r smaller)2

⇒ S bigger – S smaller = π (4)2 – π (3)2 ⇒ 16 π – 9π = 7 π

17) Answer: 7

$\sqrt{x}=2 \rightarrow x=4$

then; $\sqrt{4x}-2=\sqrt{16}-2=4-2=2$ and $\sqrt{4x+9}=\sqrt{16+9}=\sqrt{25}=5$

Then: $(\sqrt{4x+9})+(\sqrt{4x}-2)=5+2=7$

18) Answer: 12

The angles on a straight line add up to 180 degrees. Then: $3x+15+y+4y+2x=180$

Then, $5x+5y=180-15 \rightarrow 5(21)+5y=165$

$\rightarrow 5y=165-105=60 \rightarrow y=12$

19) Answer: B

$AB=5$, And $AC=12$

$BC=\sqrt{5^2+12^2}=\sqrt{25+144}=\sqrt{169}=13$

Perimeter $=5+12+13=30$; Area $=\frac{5\times12}{2}=30$

In this case, the ratio of the perimeter of the triangle to its area is: $\frac{30}{30}=1$

If the sides AB and AC become twice longer, then:

$AB=15$, And $AC=36$

$BC=\sqrt{15^2+36^2}=\sqrt{225+1{,}296}=\sqrt{1{,}521}=39$

Perimeter $=15+36+39=90$; Area $=\frac{15\times36}{2}=15\times18=270$

In this case the ratio of the perimeter of the triangle to its area is: $\frac{90}{270}=\frac{1}{3}$

20) Answer: A

The slop of line A is: $m=\frac{y_2-y_1}{x_2-x_1}=\frac{9-4}{8-3}=1$

Parallel lines have the same slope and only choice A ($y-x=8 \Rightarrow y=x+8$) has slope of 1.

Practice Test 2

Answers and Explanations

1) Answer: C

Write the proportion and solve for the missing number.

$\frac{18}{66} = \frac{3}{x} \rightarrow 18x = 3 \times 66 = 198$

$18x = 198 \rightarrow x = \frac{198}{18} = 11$

2) Answer: C

$(2^a)^b = 512 \rightarrow 2^{ab} = 512$

$512 = 2^9 \rightarrow 2^{ab} = 2^9 \rightarrow ab = 9$

3) Answer: C

If 16 balls are removed from the bag at random, there will be one ball in the bag.

The probability of choosing a white ball is 1 out of 22. Therefore, the probability of not choosing a white ball is 16 out of 22 and the probability of having not a white ball after removing 16 balls is the same.

4) Answer: B

The value of y in the x-intercept of a line is zero. Then:

$y = 0 \rightarrow -7x + 6(0) = -42 \rightarrow -7x = -42 \rightarrow x = \frac{-42}{-7} = 6$

then, x-intercept of the line is 4

5) Answer: C

Use distributive property:

$7\text{xy}(2\text{x} - 4\text{y}) = 7\text{xy}(2\text{x}) + 7\text{xy}(-4\text{y}) = 14\text{x}^2\text{y} - 28\text{x}y^2$

6) Answer: D

$y = -7x + 2$

$(-1,\ 9) \Rightarrow 9 = -7(-1) + 2 \Rightarrow 9 = 9$

$(2, -12) \Rightarrow -12 = -7(2) + 2 \Rightarrow -12 = -12$

$(-3, 23) \Rightarrow 23 = -7(-3) + 2 \Rightarrow 23 = 23$

$(2, -12) \Rightarrow 14 = -7(2) + 2 \Rightarrow 14 \neq -12$

7) Answer: B

$-41 < -6x + 1 < 13 \rightarrow$ Subtract 1 to all sides.

$-41 - 1 < -6x + 1 - 1 < 13 - 1$

$\rightarrow -42 < -6x < 12 \rightarrow$ Divide all sides by-6.(Remember that when you divide all sides of an inequality by a negative number, the inequality sing will be swapped. < becomes >)

$\frac{-42}{-6} < \frac{-6x}{-6} < \frac{12}{-6} \Rightarrow 7 > x > -2$, or $-2 < x < 7$

8) Answer: 154.

The perimeter of the trapezoid is 51 cm.

Therefore, the missing side (height) is = 51 – 13 – 9 – 15 = 14

Area of a trapezoid: $A = \frac{1}{2} h (b_1 + b_2) = \frac{1}{2} (14) (9 + 13) = 154$

9) Answer: C

A graph represents y as a function of x if

$x_1 = x_2 \rightarrow y_1 = y_2$

In choice C, for each x, we have two different values for y.

10) Answer: 21

Let x be the number. Write the equation and solve for x.

$\frac{3}{8} \times 24 = \frac{3}{7} . x \Rightarrow \frac{3 \times 24}{8} = \frac{3x}{7}$, use cross multiplication to solve for x.

$21 \times 24 = 3x \times 8 \Rightarrow 504 = 24x \Rightarrow x = 21$

11) Answer: C

To find the discount, multiply the number by (100% – rate of discount).

Therefore, for the first discount we get: $(100\% - 42\%)(E) = (0.58)$ E

For increase of 15 %:

(0.58) E$\times$ $(100\% + 15\%)$ = (0.58) (1.15) = 0.667%

12) Answer: C

When a point is reflected over x axes, the (y) coordinate of that point changes to $(-y)$ while its x coordinate remains the same.

B $(-3, 5) \rightarrow$ B' $(-3, -5)$

13) Answer: 220 m^3.

Use the volume of the triangular prism formula.

$V = \frac{1}{2}$ (length) (base) (high) $\Rightarrow V = \frac{1}{2} \times 11 \times 8 \times 5 \Rightarrow V = 220\ m^3$

14) Answer: B

Solving Systems of Equations by Elimination

$\begin{array}{l} 3x - 2y = -3 \\ \underline{7x - 5y = -6} \end{array}$ Multiply the first equation by 7, and second equation by -3, then add two equations.

$\begin{array}{l} 7(3x - 2y = -3) \\ \underline{-3(7x - 5y = -6)} \end{array} \Rightarrow \begin{array}{l} 21x - 14y = -21 \\ -21x + 15y = 18 \end{array} \Rightarrow \Rightarrow y = -21 + 18 \Rightarrow y = -3.$

$3x - 2y = -3$, $3x - 2(-3) = -3$, then: $x =- 3$, $(-3, -3)$

15) Answer: D

Use distance formula:

$C = \sqrt{(x_A - x_B)^2 + (y - y_B)^2} \Rightarrow C = \sqrt{(8 - 4)^2 + (-4 - (-1))^2}$

$C = \sqrt{(4)^2 + (-3)^2} \Rightarrow C = \sqrt{16 + 9} \Rightarrow C = \sqrt{25} = 5$

16) Answer: C

If two lines are parallel with each other, then the slope of the two lines is the same.

Then in line $y = 9x - 4$, the slope is equal to 9

And in the line $9x - y = -7 \Rightarrow y = 9x + 7$, the slope equal to 9

17) Answer: C

The sum of the weight of all girls is: $13 \times 35 = 455$ kg

The sum of the weight of all boys is: $22 \times 45 = 990$ kg

The sum of the weight of all students is: $455 + 990 = 1{,}445$ kg

average $= \frac{\text{sum of terms}}{\text{number of terms}}$; average $= \frac{1{,}445}{35} = 41.29$

18) Answer: C

$8x - 5y = 13$ has a graph that is a straight line. All other options are not equations of straight lines.

19) Answer: 20

The sum of supplement angles is 180. Let x be that angle. Therefore,

$x + 8x = 180$; $9x = 180$, divide both sides by 9: $x = 20$

20) Answer: B

The probability of choosing hearts or diamonds or spades is $\frac{39}{52} = \frac{3}{4}$

"End"